BRIDGING
SCIENTIFIC AND INDIGENOUS WAYS
OF KNOWING NATURE

Glen Aikenhead
Professor Emeritus
Aboriginal Education Research Centre
University of Saskatchewan
Saskatoon, SK, Canada

Herman Michell
Director
Northern Teacher Education Program
Northern Professional Access College
La Ronge, SK, Canada

Copyright © 2011 Pearson Canada Inc., Toronto, Ontario.

All rights reserved. This publication is protected by copyright and permission should be obtained from the publisher prior to any prohibited reproduction, storage in a retrieval system, or transmission in any form or by any means, electronic, mechanical, photocopying, recording, or likewise.

Portions of this publication may be reproduced under licence from Access Copyright, or with the express written permission of Pearson Canada Inc., or as permitted by law. Permission to reproduce material from this resource is restricted to the purchasing school.

The information and activities presented in this work have been carefully edited and reviewed. However, the publisher shall not be liable for any damages resulting, in whole or in part, from the reader's use of this material.

Permission to reprint copyright material is gratefully acknowledged. Every effort was made to trace ownership of copyright material, secure permission, and accurately acknowledge its use. For information regarding permissions, please contact the Permissions Department through www.pearsoncanada.ca.

Feedback on this publication can be sent to editorialfeedback@pearsoned.com.

Pearson Canada Inc.
26 Prince Andrew Place
Don Mills, ON M3C 2T8
Customer Service: 1-800-361-6128

ISBN-13: 978-0-13-210557-6

Vice-President, Publishing: Mark Cobham
Publisher: Reid McAlpine
Research and Communications Manager: Martin Goldberg
Managing Editors: Joanne Close, Lee Ensor
Developmental Editors: Suzanne Keeptwo, Adrienne Montgomerie, David Peebles, Eileen Pyne-Rudzik
Science Editorial Assistant: Marissa Murray
Senior Production Editor: Susan Aihoshi
Copy Editors: Susan Aihoshi, Dawn Hunter
Proofreader: Linda Szostak
Senior Production Coordinator: Louise Avery
Composition: David Cheung
Cover and Interior Design: David Cheung
Cover Image Credit: radius images/First Light Stock Photography

3 4 5 WEB 15 14 13
Printed and bound in Canada.

Contents

PREFACE ix

CHAPTER 1 Introduction 1
    Recent and Coming Changes to Science Curricula 2
    A Vignette 3
    What the Book Is About 4

CHAPTER 2 Reasons for Placing Indigenous Knowledge in School Science 7
    Equity and Social Justice 7
    Strength of a Nation's Economy 10
    Improvement of Eurocentric Science 10
    Preparation of Science-Oriented Students for Science Careers 12
    Indigenous Sovereignty and Cultural Survival 13
    Enhancement of Human Resiliency 14
    Positive Results of Integration 15
        Africa 15
        United States 15
        Aotearoa New Zealand 16
        Australia 16
        Canada 17
    Conclusion 18

CHAPTER 3 **Eurocentric Science: Background** 19
  A Quick Look at the Historical Evolution
  of Eurocentric Science 20
  Scientists: Who Are They? 22
  Other Views of Eurocentric Science 25
  Worldview 26
  The Culture of Eurocentric Science 28
  The Problem with Universalism 29
  Clarification of *Science* 30

CHAPTER 4 **Eurocentric Sciences** 33
  Diversity within Eurocentric Science.............................33
  The Myth of *the* Scientific Method................................36
  A More Realistic View of Eurocentric Sciences 37
  The Myth of Achieving Objectivity................................41
  Fundamental Presuppositions...43
    Nature Is Knowable 44
    Eurocentric Sciences Are Embedded in
    Social Contexts 44
    Eurocentric Sciences Have Predictive Validity 46
    Scientific Knowledge Is Dynamic............................47
    Scientific Knowledge Is Generalizable 48
    Eurocentric Sciences Operate on Rectilinear Time...49
    Eurocentric Sciences Subscribe to
    Cartesian Dualism 50
    Eurocentric Sciences Are Reductionist....................51
    Eurocentric Sciences Are Anthropocentric..............52
    The Material World Is Governed by Quantification ..52
    Reality Is Reproduced or Represented by
    Scientific Knowledge 54
    Empirical Data Speak for Themselves: Positivism 57
  Conclusion...59

CHAPTER 5 Indigenous Knowledge: Background.....................63

    Clarification of *Indigenous* 63

    Clarification of *Knowledge* and *Nature*..........................65

    Clarification of *Coming to Know*......................................69

CHAPTER 6 Indigenous Ways of Living in Nature 72

    Fundamental Attributes 73

        Place-Based..73

        Monist 75

        Holistic 77

        Relational 78

        Mysterious 81

        Dynamic 82

        Systematically Empirical..83

        Based on Cyclical Time..87

        Valid 88

        Rational...90

        Spiritual 92

    The Concept of Knowledge Revisited 94

    Conclusion..96

CHAPTER 7 Comparing the Two Ways of Knowing Nature 99

    Deborah's Story..100

    Hidden Pitfalls to Avoid When Comparing
Two Cultures...103

        Stereotyping..103

        Language...103

        Different Versions of Eurocentric Sciences 105

        Summary 105

    Comparisons 106

        Similarities 106

        Differences..108

A Scheme for Comparisons 112
Classroom Applications 113
Conclusion 114

CHAPTER 8 Building Bridges of Understanding: General Advice for Teachers **121**

Resources for Science Teachers 123
    Elder Involvement 123
    Community Contexts 126
    Role Models 128
    Teaching Materials and Resources 129
Indigenous Student Learning 131
Classroom Environment 133
Instructional Approaches 136
    Time-Honoured Indigenous Ideas about Teaching 137
    Other Approaches to Teaching 137
Deborah's Story Revisited 141
Student Assessment 143
Indigenous Languages 145
Teacher Expectations 148
Mr. Chang's Story 149
Conclusion 153

APPENDICES

APPENDIX A   A Cross-Cultural Science Unit 156
APPENDIX B   Questions for Reflection and Discussion 159
APPENDIX C   Website Resources 167
APPENDIX D   Recommended Books about Indigenous Worldviews 173

REFERENCES 177

INDEX 190

Acknowledgement

Nothing really dies out in a circle,
things might get old and wear away but they renew again,
generation after generation. That is what the circle is about.
—Nakawē (Saulteaux) Elder Danny Musqua

We acknowledge the future dreams and accomplishments of seven generations of Indigenous and non-Indigenous students whose science teachers have come to bridge cultures to create the circle in their classrooms.

Reviewers

DR. STEVE ALSOP
Professor, Science Education
Faculty of Education
York University
Toronto, Ontario

DR. MARLENE ATLEO
Ahousaht First Nation, British Columbia Nuu-chah-nulth Tribal Council
Associate Professor,
Coordinator, Adult and Post Secondary Education
Faculty of Education
University of Manitoba
Winnipeg, Manitoba

JEFF BAKER
Métis
Doctoral Candidate
Department of Curriculum and Pedagogy,
University of British Columbia;
Visiting Scholar
Department of Educational Foundations,
University of Saskatchewan
Saskatoon, Saskatchewan

RON BALLENTINE
Instructional Coordinator
Science and Technology, Environmental Education
Halton District School Board
Burlington, Ontario

RAY BARNHARDT
Professor of Cross-Cultural Studies
University of Alaska Fairbanks
Fairbanks, Alaska

STELLA BATES
District Principal (Ret'd)
Aboriginal Education
School District 68
Nanaimo, British Columbia

DR. MARIE BATTISTE
Mi'kmaq
Professor, Educational Foundations,
Director, Aboriginal Education Research Centre,
University of Saskatchewan
Saskatoon, Saskatchewan

TOOIE CASAVANT
Hupacasath First Nation
First Nations Educator
School District 70
Port Alberni, British Columbia

STEVEN M. DANIEL
Coordinator, Science and Secondary Education
Department of Education, Culture and Employment
Government of the Northwest Territories
Yellowknife, Northwest Territories

BRIGITTE EVERING
PhD Candidate, Indigenous Studies, Trent University;
Secondary School Teacher
Kawartha Pine Ridge DSB
Peterborough, Ontario

BARBARA FILION
Innu-Montagnais Nation, Mashteuiatsh
Royal Ontario Museum, Education Department;
Turtle Island Conservation, Toronto Zoo,
Toronto, Ontario

DEBORAH KITCHING
Science and Technology Coordinator
Fort McMurray Catholic School District
Fort McMurray, Alberta

SANDY LAST
Principal
Science Alternative Program
Langevin School
Calgary Board of Education
Calgary, Alberta

F. HENRY LICKERS
Seneca Nation Turtle Clan
Environmental Science Officer
Mohawk Council of Akwesasne
Akwesasne, Ontario

CATHERINE LITTLE
Program Coordinator
Science, Environmental and Ecological Studies
Toronto District School Board (on secondment);
Course Director,
Faculty of Education
York University
Toronto, Ontario

DR. NANCY C. MARYBOY
Cherokee and Navajo (Diné)
President, Indigenous Education Institute
Friday Harbor, WA;
Adjunct Professor,
Department of Physics and Astronomy
University of Northern Arizona
Flagstaff, Arizona

BARBARA A. MCMILLAN
Assistant Professor, Science Teacher Education
Faculty of Education
University of Manitoba
Winnipeg, Manitoba

DR. MICHAEL MICHIE
Educational Consultant
Batchelor Institute of Indigenous Tertiary Education
Darwin, Northern Territory

DR. JACQUELINE OTTMANN
Saulteaux (Nakawē)
Associate Professor
Faculty of Education
University of Calgary
Calgary, Alberta

BRAD PAROLIN
Instructional Leader
Science & Technology
Toronto District School Board
Toronto, Ontario

DR. ERMINIA PEDRETTI
Professor, Science Education
Ontario Institute for Studies in Education, University of Toronto
Toronto, Ontario

DR. ANN SHERMAN
Dean of Education
Faculty of Education
University of New Brunswick
Fredericton, New Brunswick

DR. GLORIA SNIVELY
Professor, Co-Director,
Graduate Program in Environmental and First Nations Education
Faculty of Education,
University of Victoria
Victoria, British Columbia

DR. GEORGINA MARJORIE STEWART
Maori, Ngapuhi
Lecturer, Te Puna Wananga
Faculty of Education
The University of Auckland
(Tai Tokerau Campus)
Auckland, New Zealand

TYE SWALLOW
Senior Science Instructor/Language Apprenticeship Facilitator
Saanich Adult Education Centre
W̱SÁNEĆ School Board
Victoria, British Columbia

DR. PETER CHARLES TAYLOR
Associate Professor of Transformative Education, Science and Mathematics Education Centre (SMEC)
Curtin University
Perth, Western Australia

DR. CHRISTOPHER D. TODD
Associate Professor, Biology
College of Arts and Science
University of Saskatchewan
Saskatoon, Saskatchewan

DR. PAMELA ROSE TOULOUSE
Associate Professor
School of Education
Laurentian University
Sudbury, Ontario

TED VIEW
Science Teacher
Regina Catholic Schools
Regina, Saskatchewan

PAULINE WAITI
Maori, Te Rārawa,
Manager, Māori and Pasifika Education Publications Team
Learning Media Ltd.
Wellington, New Zealand

Preface

This section gives essential information so that a reader has a sense of who the authors are and what to expect and what not to expect from the book. Each author's biography is presented from two cultural perspectives. Some general ideas about the book follow.

The Authors

Tān'si. Glen *nit'sīyikā san.* I am of British ancestry. My father, Douglas Aikenhead, was raised on a prairie farm 100 years ago and late in life became a university professor. My grandfather, John Aikenhead, grew up on a farm in the Ottawa Valley, Ontario. In the 1880s, he worked as a labourer, helping survey Saskatchewan, before settling into farming in Manitoba. His father, David Aikenhead, was raised on an inhospitable farmstead in the Ottawa Valley and became a farmer on fertile land along Ontario's Mississippi River. His father, James Aikenhead, was a farmer in Scotland. He and his family were driven off the land in 1818 by textile industrialists to make room for herds of sheep. He brought his family to the Ottawa Valley and settled on a piece of Algonkin hunting land (forested rocky marshland). I acknowledge the Algonkian Nation as inhabitants of the land that shadows my life today. I grew up in rural Alberta and urban Calgary, and I have spent the past 40 years in Saskatoon, Saskatchewan. My white, male, middle-class identity has always provided me with a privileged status in Canada. From my ancestors I have learned that with privilege come responsibilities, such as to people historically marginalized in school science. *Ekosi.*

Glen Aikenhead is Professor Emeritus, University of Saskatchewan, where he began working in 1971. He earned an Honours Bachelor of Science (University of Calgary, 1965), Master of Arts in Teaching (Harvard University, 1966), and a Doctorate in Science Education (Harvard University, 1972). He taught secondary school science in Calgary and internationally in Germany and Switzerland. Glen has always embraced a humanistic perspective on science, even as a young research chemist in Canada. This perspective guided his educational research and development over the years in such academic fields as curriculum policy, student assessment, development of classroom materials, and classroom instruction. His university teaching included science methods courses for First Nations and Métis teacher education programs. This work led him into the field of culture studies in science education in the early 1990s, with First Nations and Métis peoples of Canada and with colleagues worldwide. He has published widely in his fields of interest, including, in 2000, a community-based, teacher-collaborative project titled *Rekindling Traditions* (six cross-cultural science units for Indigenous students in Grades 6 to 11).

I am Herman Michell, a Woodlands Cree (*Nîhîthewâk*) whose ancestry includes an influence of Dëne and a sprinkle of Inuit and Swedish. I am fortunate to speak my Cree language ("th" dialect). I grew up in the tiny Woodlands Cree trapping and fishing community of Kinoosao, on the eastern shores of beautiful Reindeer Lake in northern Saskatchewan, at the Manitoba border. The Michell family comes from a long history of hunters and trappers, going back hundreds of years. My father, John Michell, was a product of intermarriages between the Dëne people who occupied the Reindeer Lake area and Woods Cree families. His parents were Apikosis (Mouse) and Tawipisim (Half Sun). He spoke three languages: Cree, Dëne, and English. My mother, Therese Peterson, was born in Yellowknife, Northwest Territories, to an Inuit mother, Carlston Inuik, and Swedish father, Alfred Peterson, a trapper who became acquainted with my father. My first teachers of the natural world were my parents, who showed me the ways of the land and the importance of bringing this knowledge into the education system. We lived a sustainable and nomadic lifestyle, moving according to the seasons, cycles, and movements of the animals we depended on for survival.

Herman Michell is Director of the Northern Teacher Education Program, Northern Professional Access College, La Ronge, Saskatchewan. Formerly, he was an Associate Professor at the First Nations University of Canada, where he taught and served his community for more than 10 years. He earned a Bachelor of Arts (University of Winnipeg, 1990), completed Post-Baccalaureate course work in Educational Psychology (University of Manitoba, 1994), and earned a Master of Education (University of British Columbia, 1998), a Certificate in Teaching in Higher Education (University of Regina, 2000), and a Doctorate in Education (University of Regina, 2008). His vision is to build cultural bridges between First Nations peoples and Western science to increase the numbers of his people in science- and health-related careers. With expertise in curriculum and instruction and a focus on First Nations cultural content inclusion, Herman has published widely in academic journals and has travelled internationally. He dedicates his time to promoting science literacy in First Nations communities by engaging in community projects and by talking with teacher candidates in science education. As principal investigator, he and his research team completed a major federally funded study in 2008: "Learning Indigenous Science from Place: An Action Research Study Examining Indigenous Science Perspectives in Saskatchewan First Nations and Métis Community Contexts." It will influence curricula and instruction for years to come.

The Book

The recognition of Indigenous knowledge[1] as an important and legitimate source of understanding of the physical world is increasing within education jurisdictions worldwide. Consequently, this content appears, or is beginning to appear, in science curricula. Science teachers, teacher candidates in university programs, and school administrators who face the challenge of implementing such a curriculum (especially in Grades 6 to 12) are the primary audience for this book. It will also be valuable to university Native Studies programs, ministry of education personnel, and policy-makers.

To build cultural bridges between Indigenous and scientific ways of knowing nature, one first needs a contemporary and general understanding of each knowledge system. This understanding is the book's first emphasis: It offers insights that help teachers construct their own understanding of how both knowledge systems describe and explain nature. It explores commonalities and differences between the systems.

A second emphasis concerns the practical aspects of interpreting or teaching some of the Indigenous knowledge found in a science curriculum. To this end, we speak initially from a Saskatchewan experience (e.g., Indigenous spelling and word usage), since Saskatchewan, after the territories, is the province with the second-highest proportion of Indigenous citizens in Canada (14.9%) (Richards & Scott, 2009, p. 6). We also draw heavily

---

[1] This and other terms are defined in Chapter 1.

on the personal experiences of Indigenous educators and writers to make this book's content applicable to other regions of Canada and to other places with Indigenous populations, such as the United States, Australia, Aotearoa New Zealand, South America, Taiwan, and northern Europe.

Teaching opportunities that involve Indigenous knowledge that are discussed in our book usually originate in rural settings, not in cities where Indigenous students may feel particularly isolated from their rural roots. Nevertheless, the book speaks to the formation and strengthening of Indigenous identities for rural or urban students, in keeping with students' ancestry. A recent study of 11 cities across Canada concluded, "Urban First Nations peoples, Métis and Inuit alike maintain great reverence for their heritage and express strong Indigenous pride" (Environics Institute, 2010, p. 4). This fact could be a positive motivation in urban science classrooms; however, we acknowledge that more resources are available for teachers in rural settings.

The implications for the identities of non-Indigenous students in a multicultural classroom are considered. The book encourages urban teachers to transfer ideas from the text to create cultural connections for their students, but the book does not address specific challenges of urban youth in multicultural settings.

Reading a book is not adequate for understanding specific Indigenous practices (e.g., berry picking or fishing), which invariably require experiential learning. This book is, however, an excellent place to begin *appreciating* what both contemporary and traditional Indigenous knowledge mean. This appreciation prepares science teachers to engage more effectively with an Indigenous community, which is the prime source of specific Indigenous knowledge available to teachers.

Our book cannot be all things to all readers. For instance, it is not a recipe for what to do Monday morning. But it does offer knowledgeable *perspectives* on scientific and Indigenous content. It does serve as a practical guide for *approaching the decision* of what to do Monday morning. However, specific content and instruction methods for Monday morning will vary considerably from community to community.

CHAPTER 1

Introduction

One aim of school science is to enrich all students' lives by conveying how academic scientists understand nature. Some students enjoy understanding their physical world in a way similar to their science teachers. They share a scientific worldview.[2] These science-oriented students eagerly acquire a scientific mindset as they learn the standards of a scientific discipline. Science-oriented students want to think, talk, and believe the way academic scientists do. Some will eventually become science teachers, scientists, or engineers.

However, not all students possess such a scientific worldview. Empirical research shows that a majority of students[3] prefers to understand nature through other worldviews (Aikenhead, 2006), such as primarily aesthetic, religious, or economically pragmatic ones (Cobern, 2000). These science-shy students tend to be much less enthusiastic about thinking, talking, and believing scientifically. On a personal level, they do not relate easily to a scientific worldview and they often experience school science as a foreign culture. Science-shy students become uninterested and can be frustrated or even alienated by their experiences (Taconis & Kessels, 2009). Their intuitive or subconscious reaction may be so subtle that science teachers seldom detect it (Aikenhead, 2006). In spite of their best intentions to enrich all students' lives through science, many science teachers may not be aware of their students' adverse feelings toward school science, unless the teachers are acquainted with students' worldviews that differ so much from scientific worldviews.

---

[2] In general, a person's worldview determines the way he or she experiences and makes sense of the world. This topic is discussed in Chapter 3, in the section "Worldview."

[3] In North America, this group comprises about 90% of the student population in a typical high school (Aikenhead & Elliott, 2010).

One response to this dilemma has been a science-technology-society-environment (STSE) approach to school science, illustrated by the *Pan-Canadian Common Framework for Science Learning Outcomes* (Council of Ministers of Education Canada, 1997). An STSE approach embeds scientific content in an everyday context to create a need to know that content in students' minds. This approach gives students a more relevant experience that focuses on meaningful learning. It enables them to understand concepts well enough to make sense of phenomena rather than simply memorizing vocabulary, definitions, and algorithms (Hutchison & Hammer, 2010; Pedretti & Little, 2008). Worldwide, STSE curricula have met with documented success over decades of implementation (Aikenhead, 2006). This book builds on those successes.

But success in school science continues to elude most Indigenous[4] students, including First Nations, Inuit, and Métis students in Canada; American Indian students in the United States; Māori students in Aotearoa New Zealand; students from the many Aboriginal families who live throughout Australia; Quechua students in Peru; Sámi students in northern Europe; and students of Aboriginal tribes in Taiwan. Achievement in school science remains a major challenge. Indigenous students are underrepresented in high school science courses and university science-related programs.

Recent and Coming Changes to Science Curricula

The last two decades of the twentieth century witnessed an international renaissance in Indigenous cultures.[5] Indigenous peoples began to assert their human rights and sovereignty (Niezen, 2003). In 1992, the Aotearoa

---

[4] The term *Indigenous*, defined in detail in Chapter 5, encompasses worldwide the original inhabitants of a place and their descendants who have suffered colonization. The term includes Canada's Aboriginal peoples, which in the Canadian Constitution refers collectively to the First Nations, Inuit, and Métis peoples of Canada.

When referring to more than one Indigenous group, the term *Indigenous peoples* applies; when referring to more than one Indigenous person, no matter what cultural groups are involved, the term *Indigenous people* applies. This convention is used throughout the book.

[5] In general, *culture* means "a blanket of comfort that gives meaning to lives" (Davis, 2009, p. 198). More specifically we use the word to refer to the norms, values, beliefs, knowledge, language, technologies, expectations, and conventional actions shared by a community as large as a nation and as small as a village.

New Zealand science curriculum was translated into Māori by Māori Elders and taught in designated schools (McKinley, 1996). Some jurisdictions in the United States have placed Indigenous knowledge in their science curricula (Cajete, 1999; James, 2001; Kawagley, Norris-Tull, & Norris-Tull, 1998). Australian national curriculum policies explicitly support including "Indigenous science" (Michie, 2002).

In the spirit of reconciliation, the twenty-first century has seen a number of ministries and departments of education in Canada recognize Indigenous ways of knowing nature as fundamental content in school science.[6] (Chapter 2 summarizes reasons for this.) With the guidance of First Nations, Inuit, and Métis communities in Canada, each province and territory determines what Indigenous knowledge will appear in its science curriculum. Conventional science content will continue to be taught, but it will no longer be presented as the *only* legitimate way to understand nature.

A Vignette

Ms. Kirsten Smith, a Grade 8 science teacher with three years of experience, hoped to motivate the Indigenous students in her class by making science more relevant to them. Her principal had encouraged all teachers to include culturally responsive instruction in their lessons where possible, instruction that would include both content and teaching methods that respect and draw on a student's cultural community. Ms. Smith decided to try out a unit on snowshoes. An Internet search found her a set of lesson plans developed by teachers who had collaborated with members of their Indigenous community. She knew the physical education teacher had a class set of snowshoes she would need for the lessons. But teaching a science unit that included Indigenous knowledge was very new to her.

Ms. Smith followed the advice in the lesson plans and advised parents about what new things would be happening in science class. Through an Indigenous bank teller she knew, Ms. Smith met Jim, an Indigenous retired farmer who had made snowshoes the "traditional way" when he was young. Her first visit to his home to learn more about snowshoes opened up a

---

[6] These ways of knowing nature are also known as Indigenous knowledge, Aboriginal knowledge, Aboriginal science, Indigenous science, traditional knowledge, traditional ecological knowledge, Native science, and so on. This book uses *Indigenous knowledge*, but other terms appear in the literature.

new world. She felt inspired to continue with her teaching plans but felt inadequate not knowing more about the following points:
- What kind of knowledge is this new content?
- What does it have in common with the knowledge currently taught in science classes?
- What are some important differences a teacher should know about?
- How can a teacher better prepare to implement this culturally responsive teaching?

What the Book Is About

The purpose of this book is to provide answers to Ms. Smith's questions. The text contains academic references so ideas may be further explored. The implementation of a science curriculum enhanced by Indigenous knowledge requires professional development and support for teachers. This book is primarily for science teachers and teacher candidates faced with the challenge of teaching the Indigenous knowledge identified in their science curriculum. It can help the reader prepare a professional development program, become part of that program, and serve as a resource for years to come. Teachers in urban classrooms can use the book to extend a culturally sensitive hand to urban Indigenous students, as well as non-Indigenous multi-ethnic students.

Because content and methods vary considerably across regions, specific questions on what details of Indigenous knowledge should be taught, and the methods used to teach them, must be answered by a ministry or department of education and the school's local Indigenous community. Those details are outside the scope of this publication.

But, to help such people as Ms. Smith better understand the enhanced Indigenous curriculum content to be taught, this book summarizes and then compares two different ways of knowing nature—*Eurocentric science*[7] and *Indigenous knowledge* (Indigenous ways of knowing nature). Because these two labels obscure the great diversity found within each group, they can unfortunately lead to stereotyping. This book aims to counter such

---

[7] For the sake of clarity and accuracy, we will refer to science as *Eurocentric science* in this book because science conveys Eurocentric ways of understanding the universe, ways that were originally developed in Eurocentric cultures, as described in Chapter 3. Science is also called *Western science* for the same reason.

stereotyping of these two ways of knowing nature found in the media and elsewhere.

The labels also mask important similarities between the two groups. For example, both knowledge systems developed from culture-based ways of experiencing and making sense of nature—primarily Euro-American cultures for Eurocentric science, and Indigenous cultures for Indigenous knowledge. Furthermore, each knowledge system *in its own cultural way* relies on empirical data, experimental types of procedures, rationality, intuition, and predictability.

We try to avoid using an either/or dichotomy—Eurocentric science *versus* Indigenous knowledge—because such contrast often subliminally conveys a superior–inferior relationship. Historically, the dichotomy has denied or marginalized Indigenous peoples' points of view. Instead of either/or, we discuss how *both* ways of knowing can enrich our understanding of nature, a position supported by a number of ministries and departments of education in Canada and internationally. In short, Eurocentric science and Indigenous knowledge represent complementary, not separate, realities. The two can coexist.

This book will help science teachers construct personal points of view about how each knowledge system describes and explains nature. Elders do not ask science teachers to embrace their Indigenous knowledge but to acknowledge and respect their different, yet legitimate, ways of knowing nature.

Later, it will become clear that the terms *Eurocentric science* and *Indigenous knowledge* are not quite accurate and that they can be replaced by other terms that better describe each type of knowledge system.[8] More precise terms will improve science lessons or units that involve both Eurocentric science and Indigenous knowledge. These science units, sometimes called *cross-cultural science units*, will engage students in two different but complementary ways of knowing nature. (Appendix A describes an example of a unit that combines Eurocentric science and Indigenous knowledge.) Science curricula will typically determine the number of times a cross-cultural science lesson or unit is taught.

By exploring fundamental commonalities and differences between Eurocentric science and Indigenous knowledge, this book provides insights for science teachers to build bridges between the familiar scientific ways of

---

[8] These other terms are *Eurocentric sciences* and *Indigenous ways of living in nature*.

knowing nature and the less familiar Indigenous ways. We can be inspired by Marie Battiste (2000), a Mi'kmaw scholar and recipient of a 2008 Canadian Aboriginal Achievement Award. She wants to find ways of "healing and rebuilding our nations ... by restoring Indigenous ecologies, consciousnesses, and languages and *by creating bridges* between Indigenous and Eurocentric knowledge" (p. xvii, emphasis added). Yupiaq science educator Oscar Kawagley (1990) in Alaska contends that strong bridges are built by examining the collective ways people in Eurocentric and Indigenous cultures experience and make sense of their natural worlds—which is exactly what this book is about.

For science teachers of students whose cultural ancestry is neither European nor Indigenous, the book also supports bridge-building techniques to other cultures, whenever feasible.

The book has eight chapters. Chapter 2 summarizes the reasons why Indigenous knowledge should be included in school science. Chapters 3 and 4 clarify what Eurocentric science is and provide a basis for comparing it with Indigenous knowledge. These two chapters invite teachers to reflect on fundamental concepts of Eurocentric science, such as what reality is, how scientists know the world and what their ideas are based on, as well as the values that guide their work. Our objective is to present a realistic account of Eurocentric science in everyday life rather than a stereotype. Similarly, Chapters 5 and 6 offer an overview of fundamental attributes that characterize Indigenous knowledge—ideas about what reality is, how Indigenous peoples know the world and what their ideas are based on, as well as the values that guide Indigenous communities. These parallel topics allow comparisons of the two knowledge systems, thus supporting teachers' efforts in bridging the cultures. Comparisons are summarized in Chapter 7, which includes a story of an Indigenous science student dealing with cross-cultural challenges in biology. Chapter 8 offers practical advice and general suggestions on teacher resources, student learning, classroom environment, instructional approaches, student assessment, the importance of Indigenous languages, and teachers' expectations of Indigenous students. It ends with a story from an Indigenous cross-cultural science classroom. Note: Because the advice and suggestions in Chapter 8 were composed in the context of Chapters 3 to 7, misunderstandings may arise if Chapter 8 is read in isolation from Chapters 3 to 7.

CHAPTER 2

Reasons for Placing Indigenous Knowledge in School Science

There are several reasons for integrating Indigenous knowledge into the school science curriculum. This chapter looks at these reasons and provides evidence of successful integration.

Equity and Social Justice

In the 1980s, science educators as well as ministries and departments of education became concerned with the under-representation of women in science-related careers. A principal reason for this under-representation is that young women's cultural self-identities did not always harmonize with the male-dominated culture of Eurocentric science (Brotman & Moore, 2008; Organisation for Economic Co-operation and Development [OECD], 2006b; Science Council of Canada, 1984). Other reasons included societal expectations, such as family responsibilities associated with child-bearing. Since then, an awareness of equity and social justice, along with a need for a larger scientific labour pool, have inspired innovations by science educators that address the culture of school science as applied to gender differences.

Today, we face another challenge. "There is abundant evidence that Aboriginal people are under-represented in science and technology occupations and educational programs" (Canadian Council on Learning, 2007b, p. 4). A lack of representation of Indigenous peoples is evident in the science and engineering professions as well as in many science-related occupations such as healthcare workers, managers, technicians, and employees in resource industries (Battiste, 2002; MacIvor, 1995). The *under-representation* of Indigenous students in high school science courses is a major challenge for science educators in industrial countries.

The lack of Indigenous peoples in scientific fields results in economic and social disadvantages for Indigenous communities (MacIvor, 1995; Royal Commission on Aboriginal Peoples [RCAP], 1996). To address these disadvantages and the ethical problems they create, many ministries and departments of education have revised or soon will be revising science education policies for more equitable representation of Indigenous students in school science to support their success (Richards & Scott, 2009).

The issue of equity and social justice is complex because many circumstances influence this under-representation. Undermining an Indigenous family's support for their child's successful education are generations of colonial oppression such as residential schools and the Indian Act, the presence of racism, systemic poverty, chronic underfunding by federal governments, and continued adverse living conditions. Science educators have no direct influence over these factors.

However, science teachers can influence Indigenous students' marginalization in science classrooms and in other school subjects on the basis of students' cultural identities (Battiste, 2002). One way to understand this finding is to acknowledge the clashes that many Indigenous students experience between their own culture and that of Eurocentric science embedded in school science (Aikenhead, 2006; Akatugba & Wallace, 2009). Values and assumptions embraced by Eurocentric science (see Chapters 3 and 4) often conflict with those expressed in Indigenous ways of knowing and experiencing nature (see Chapters 5 and 6). This cultural difference may be expressed as follows: The way scientists *see* the world can clash with the way Indigenous Elders *inhabit* the world. According to the Canadian Council on Learning (2007b), this type of cultural conflict causes many Indigenous students to feel that school science is too foreign to them and, thus, irrelevant. Eurocentric science's technical language and impersonal style can be alienating (Hodson, 2009, Ch. 8). They include the didactic, formal presentation of scientific concepts and the need to make distinctions between concepts not existing in students' language, for example, asexual and sexual reproduction.

Moreover, school science instruction can be experienced by Indigenous students as an attempt to assimilate them into the dominant culture. Many Indigenous students do not relate to or may resist Eurocentric science passively or actively. Subsequently they fail to enrol or succeed in science courses. School science has historically excluded Indigenous knowledge from the curriculum. Students' resistance to school science is often lessened with culturally responsive teaching (see Chapter 8), which attempts to diminish the effects of colonization.

Indigenous students' marginalization in science education has long been a concern to Indigenous leaders; ministries and departments of education; federal governments; many universities; as well as scientific organizations such as the American Association for the Advancement of Science (AAAS) (1977), the International Council for Science (2002), and the Science Council of Canada (1991). Although thoughtful innovations and interventions have successfully encouraged some science-oriented Indigenous students to participate in science courses and programs, these initiatives fall short of resolving the general problem of under-representation. Only when Indigenous students are offered a school science experience that supports their cultural identity and respects their inherent ways of knowing nature can they perceive themselves as valid participants in science-related careers and occupations (Cajete, 1999; Hunt & Harrington, 2008).

According to the Council of Ministers of Education Canada (2002) and many other educational jurisdictions worldwide (McKinley, 2007), a key to improving enrolment and retention rates of Indigenous students is to offer an enhanced science curriculum that recognizes Indigenous knowledge as a foundation for understanding their place in the world. In such a curriculum, some Indigenous knowledge will become as intrinsic as the concepts of energy, biomes, and chemical equilibrium. For many Indigenous students, an enhanced science curriculum offers significant and relevant content, making school science feel less foreign.

In Canada, each province and territory will determine the balance between Indigenous knowledge and topics in Eurocentric science appropriate to school science. Some provinces and territories have already done so (Aikenhead & Elliott, 2010). Whatever the balance is, an enhanced science curriculum allows Indigenous students to form strong cultural self-identities and to fully participate in society. Besides finding employment, students' goals include becoming environmentally aware, being scientifically and technologically knowledgeable, and developing a voice of conscience regarding sustainability (Shope, 1998).

Strength of a Nation's Economy

On economic grounds alone, the educational success of Indigenous students will have direct consequences for the well-being of any society. Better educational outcomes lead to increases in Indigenous people's earning power, increases in government tax revenue, and decreases in government program expenditures (Sharpe & Arsenault, 2009). When Statistics Canada (2008) released its *Aboriginal Peoples* report, the news media were quick to understand its significance. The Canadian Broadcasting Corporation (CBC) (2008) reported that Canada's Aboriginal population surged past the one million mark for the first time on a census, a spike of 45% from a decade earlier. Newspapers across Canada published articles and editorials devoted to the consequences of this increase in population. An editorial written by the Saskatoon *StarPhoenix* ("Editorial: Poor literacy," 2008) stated, "Those who argue that the status quo of forcing the Aboriginals to adapt to the existing system will work, are merely burying their heads in the sand and damaging the economy, as were those in the past who argued the way to educate the Natives was to destroy their culture."

Education—"the new buffalo" (Stonechild, 2007)—is seen by the media and policy-makers (Richards & Scott, 2009) as a major contributor to economic progress for a country. On a NASA website, Lakota Chief Joseph Chasing Horse stated, "We once hunted for buffalo, we now hunt for knowledge" (quoted in Shope, 1998, p. 7). Complicity with the status quo is no longer a viable option.

An enhanced science curriculum that explicitly supports both Indigenous knowledge and scientific literacy moves science education in an economically favourable direction. It also broadens school science literacy to become *knowing-nature literacy*.

Improvement of Eurocentric Science

Some scientists want to expand the content of Eurocentric science to encompass ideas from Indigenous knowledge to improve the contributions of Eurocentric science to our planet's sustainable future (Clark & Dickson, 2003; Snively & Corsiglia, 2001). The International Council for Science (2002) recognizes the importance of including "traditional knowledge" within its organization. Knudtson and Suzuki (1992, p. xxiv) argue, "We must create for ourselves a sense of place within the biosphere that is steeped in humility and reverence for all other life." They deplore situations in which government officials and scientists make crucial decisions over resource management while ignoring highly relevant Indigenous knowledge. Such poor decisions have had devastating consequences for some communities (McGregor, 2000; Nadasdy, 1999). Alternatively, there are scientists such as Henry Lickers, who is a member of the Turtle Clan of the Seneca Nation, a biologist, and the Director of the Department of the Environment for the Mohawk Council of Akwesasne in Canada. During his long career, he has been instrumental in incorporating Indigenous knowledge into environmental planning and decision making, for which he has won several achievement awards. His work demonstrates how environmental biology can be improved by Indigenous knowledge (FreshWater Summit, 2010). The past few decades have seen an increasing number of biologists, ecologists, and geologists use Indigenous knowledge learned from Elders to develop emerging scientific fields, such as "traditional ecological knowledge and wisdom" (TEKW) (Menzies, 2006; Turner, Ignace, & Ignace, 2000).

Unfortunately, Indigenous terrestrial and marine environments, along with their genetic resources, have been pillaged by globalization generally and threatened by biopiracy[9] specifically, with the involvement of some corporate scientists (Settee, 2000; Sillitoe, 2007). However, enlightened businesses are embracing sustainability instead. Some Canadian companies, such as Snow & Associates in Calgary, hire Indigenous Elders to develop accurate and complete environmental impact studies. "People who depend on local resources for their livelihood are often able to access the true costs and benefits of development better than any evaluator from the outside" (Snively & Corsiglia, 2001, p. 18).

Sustainability is inherent to Indigenous knowledge, a concept not usually integral to Eurocentric science. The planet's environmental crises cannot be solved solely with conventional Eurocentric science and technology but must call on knowledge systems that naturally embrace sustainability at their very core.

An improvement in Indigenous students' interest and achievement in science courses and university programs will produce more Indigenous scientists, engineers, technicians, and health professionals. Resource management, the

---

[9] Biopiracy is an act of stealing genetic resources or Indigenous knowledge about the biological environment (e.g., the healing effects of certain plants) and unfairly profiting from their commercialization. Biopiracy often involves a corporation taking out a patent and developing a commercial product. By international rules of commerce, the corporation owns that genetic knowledge; the community from which it was taken has no ownership rights.

health sector, and research and development (R&D) in business and industry will then function with individuals unrestricted by a conventional Eurocentric mindset of scientific thinking. The collective Indigenous values, ideologies, and intuition hold promise for sustainability, biodiversity, and holistic creativity (Cajete, 2000a; International Council of Science, 2002; Settee, 2000).

Lillian Dyck (1998), a neuroscientist of Plains Cree ancestry and a senator in the Canadian Parliament, talks about how an Indigenous way of making sense of the world "can, in fact, affect scientific practice" (p. 88). The ecological health of our planet is critical here, as Viergever (1999) explains:

> What matters in the long term is the continuation of an [Indigenous] system that has shown to be able to generate knowledge…; a system that has developed alternative solutions for several local problems. Perhaps these solutions are not as "sophisticated" as the solutions developed by the scientific system, but often they are equally effective and environmentally more sustainable. (pp. 334–335)

Non-Indigenous students planning a science-related career will also benefit from learning some Indigenous knowledge. Their perspectives on nature and their creative problem-solving capabilities will be enhanced. They may become more well-rounded and reflective scientists, engineers, resource managers, and health professionals in the future. "Indigenous knowledge fills the ethical and knowledge gaps in Eurocentric education, research, and scholarship" (Battiste, 2002, p. 5). An enhanced science curriculum is beneficial for non-Indigenous students.

Preparation of Science-Oriented Students for Science Careers

Some ideas in science curricula are outdated, although technicians, engineers, and scientists continue to use them because the ideas work well enough. The modern replacements for these ideas tend to be abstract and counterintuitive. They can be both difficult and unnecessary to learn. For example, Newtonian physics relies on an idea called absolute space and time in which each is measured separately and independently. When Einstein's theories about relativity became accepted scientific knowledge, absolute space and time was replaced by a space-time continuum (Greene, 1999). According to Native American Pueblo physicist Phil Duran (2007), such sophisticated ideas as relativity, quantum theory, string theory, and chaos theory seem to be based on assumptions similar to those found in Indigenous knowledge. By teaching Indigenous knowledge in science classes, these assumptions could be introduced to science-oriented students in preparation for learning the scientific concepts at a later date.

Moreover, by learning about the similarities and differences in Indigenous knowledge and Eurocentric science, science-oriented students can better understand first, the importance of social and cultural contexts in the production of scientific knowledge (see Chapter 3); second, the type of truth associated with scientists' knowledge (see Chapter 4, the section "Eurocentric Sciences Have Predictive Validity"); and third, the strengths and limitations of Eurocentric science in general (see Chapter 4). In short, science-oriented students will gain substantially in their understanding of the nature of science (Aikenhead, 2006).

Indigenous Sovereignty and Cultural Survival

A worldwide renaissance now supports the sovereignty and cultural survival of Indigenous peoples (McKinley, 2007; Niezen, 2003). This movement toward sovereignty is about healing and rebuilding Indigenous nations oppressed by colonization: "Our defeat was always implicit in the history of others; our wealth has always generated our poverty by nourishing the prosperity of others, the empires and their native overseers" (Galeano, 1973, p. 5). The movement toward sovereignty also seeks to resolve the tension between Indigenous peoples fighting globalization and biopiracy (described in a footnote above), and companies developing new products useful to humankind along with the associated corporate profits. All countries with Indigenous citizens have a role in the worldwide movement toward sovereignty. Science education can make tangible contributions to this movement locally, nationally, and internationally.

Indigenous scientists, engineers, technicians, resource managers, science teachers, and health professionals can serve Indigenous communities by initiating economic development projects and by taking greater control of land use, resources, education, and health needs—thus enhancing Indigenous sovereignty (Michell, 2007). As Métis science educator Madeleine MacIvor (1995) points out:

> Reasserting authority in areas of economic development and health care requires community expertise in science and technology.... We [can] trans-

form the science curriculum from one which is essentially assimilationist to one which honours, respects, and nurtures our traditional beliefs and lifeways, and which presents science and technology in a more authentic way." (pp. 74, 90)

Most Indigenous families acknowledge the need to support the success of their children, both in school science and in the survival of their culture. In the words of Canada's Royal Commission on Aboriginal Peoples (RCAP, 1996, vol. 3, sec. 5, p. 2), "Aboriginal people rightly expect education to serve as a vehicle for cultural and economic renewal. But this will not happen without critical changes in education processes and systems."

Enhancement of Human Resiliency

The survival of Indigenous cultures assists all non-Indigenous peoples to deal with the accumulated effects of non-sustainable human progress that have violated our planet's life-giving biological support systems. World-renowned anthropologist Wade Davis (2009) documents the creative problem-solving resiliency of Indigenous peoples the world over, which gives hope for building bridges between cultures:

> These [Indigenous] voices matter because they can still be heard to remind us that there are indeed alternatives, other ways of orienting human beings in social, spiritual, and ecological space. This is not to suggest naively that we abandon everything and attempt to mimic the ways of non-industrial societies, or that any culture be asked to forfeit its right to benefit from the genius of technology. It is rather to draw inspiration and comfort from the fact that the path we have taken is not the only one available, that our destiny therefore is not indelibly written in a set of choices that demonstrably and scientifically have proven not to be wise. (pp. 217–218)

All global citizens should be alarmed that the environment is being destroyed worldwide. Indigenous knowledge and practices are inherently based on sustainability; therefore, knowing Indigenous knowledge and practices will encourage humanity's stewardship of Earth. "All sources of knowledge must work together as we face the implications of our global and ecological interdependence, for seven generations to come" (Shope, 1998, p. 8). Multiple ways of understanding the environment encourages "two-way learners" to create knowledge hybridized from Indigenous and Eurocentric knowledge systems and to take sustainable action (van Eijck & Roth, 2007; Sillitoe, 2007).

Positive Results of Integration

The last reason for introducing Indigenous knowledge into school science is that such integration does work, if adequate preparation and support for teachers are provided. Research has evaluated the impact of implementing an enhanced science curriculum in a culturally responsive way (Aikenhead, 2006). In these studies, cultural validity was assured by having Indigenous groups decide the content of the local science curriculum. A sampling of cross-cultural projects is offered here.

Africa

When Eurocentric science content was embedded in local African cultures, student interest and achievement increased (Jegede & Okebukola, 1991; Lubben & Campbell, 1996). As a result, post-apartheid South Africa established the goal of teaching local African knowledge of nature in science classrooms (Keane, 2008). Research in other regions of the world supports this approach.

United States

The Alaska Native Knowledge Network (ANKN, 1996) produced cross-cultural teaching materials for Yupiaq students. In classrooms where teachers implemented the ANKN science modules, the standardized science test scores of Yupiaq students uniformly improved over four years to match the national averages in the United States (Barnhardt, Kawagley, & Hill, 2000).

Chinn (2007, 2008) drew on Indigenous Hawaiian knowledge to develop an environmental literacy program for K–12 science curricula that met the standards-based expectations of the Hawaiian Ministry of Education. The program's success was partially attributed to in-depth professional development for science teachers that combined cultural immersion with formal study at the University of Hawai'i.

A review of research in other parts of the United States reached the following conclusion: "Efforts at culturally responsive schooling for Indigenous youth result in students who have enhanced self-esteem, develop healthy [cultural identities], are more self-directed and politically active, give more respect to tribal elders, have a positive influence in their tribal communities, exhibit more positive classroom behaviour and engagement, and achieve academically at higher rates" (Brayboy & Castagno, 2008, p. 733). Elementary science classes similarly experience success (Matthews & Smith, 1994); this result is also seen in multicultural classrooms (Lee, 2002).

When culturally responsive instruction included outdoor, hands-on science instruction with group activities, instead of textbook-based indoor instruction, standardized science test scores of Indigenous students improved significantly, reaching par with the scores of their non-Indigenous counterparts (James, 2001; Zwick & Miller, 1996). Similarly, Riggs (2005) concluded that "successful Earth science curricula for Indigenous learners share in common an explicit emphasis on outdoor education, a place-based and problem-based structure, and the explicit inclusion of traditional Indigenous knowledge in the instruction" (p. 296). Student success also increases whenever an Indigenous community becomes involved with enhancing its school science program (Reyhner, 2006).

Native American cross-cultural science education projects are numerous but not often publicized. Thankfully, the NASA website "Ancient Observations, Timeless Knowledge" (NASA, 2005) serves as a clearing house. The Indigenous Education Institute (2009) website also lists major projects. (See Appendix C for additional helpful websites.)

Aotearoa New Zealand

Aotearoa New Zealand has enjoyed an enhanced science curriculum for several years. A Mäori-language version of their country's standard science curriculum came about in 1992 through negotiations between Mäori Elders and science educators (McKinley, 1996, 2007; Stewart, 2005). This version was then introduced into a network of Mäori bilingual and immersion classrooms in both elementary and high schools (Wood & Lewthwaite, 2008). A thorough evaluation of these programs documented various advantages for Mäori students (McKinley, Stewart, & Richards, 2004). Aotearoa New Zealand is the most advanced country in the world in cross-cultural science teaching.

Australia

National policies in Australia explicitly support Indigenous knowledge in school science (Michie, 2002). This support inspired a science educator to collaborate closely with three different Indigenous family groups to produce a high school textbook called *The Kormilda Science Project* (Read, 2002). A different approach resulted in "Australian Indigenous Science" as Chapter 1 in a junior secondary textbook, *Science Edge 3* (Sharwood & Khun, 2005). PrimaryConnections, an Indigenous elementary school project, was enthusiastically received by teachers (Bull, 2008). The web-based

Indigenous Science Network Bulletin (see Appendix C) keeps up to date with many other similar developments.

Canada

For Canadian curricula enhanced with the inclusion of Indigenous knowledge, increased student interest and achievement in science and other school subjects have been documented (Kanu, 2002; Snively, 1990). In the far north, the Inuit Subject Advisory Committee (1996) published *Inuuqatigiit: The Curriculum from the Inuit Perspective*, which conveys Inuit Elders' ideas to be included in school science. A case study in a Grade 7 science classroom identified advantages and challenges when integration was solely the responsibility of the teacher (DeMerchant, 2002). In Nunavut, Inuit ways of knowing nature are compiled in the document *Inuit Qaujimajatuqangit*, developed conjointly by the Nunavut government and its citizens (Lewthwaite & McMillan, 2007; Department of Human Resources, 2005).

On the west coast, *Forests and Oceans for the Future* (Menzies, 2003) combined Indigenous knowledge and Eurocentric science by developing teaching materials that link the community's Indigenous knowledge with emergent anthropological research (Ignas, 2004). At the University of Victoria, a major professional development project encourages teachers' familiarity with Indigenous knowledge and the teaching of it in their classrooms (Snively & Williams, 2008). A five-year study with 366 public schools in British Columbia found that Indigenous students increased their achievement when Indigenous content was incorporated into the curriculum (Richards, Hove, & Afolabi, 2008). This research also resulted in "improving relations with Aboriginal families and community members, and transforming expectations in schools" (p. 14).

The project, *Rekindling Traditions* (illustrated in Appendix A), integrated Eurocentric science into the Indigenous knowledge of northern Saskatchewan communities (Aikenhead, 2000, 2002). The units tend to engage Indigenous students who might otherwise resist school science. Similarly, Woodlands Cree teachers in northern Manitoba developed their own culturally responsive science lessons with good results (Sutherland & Tays, 2004). On the basis of several years of work, Sutherland and Dennick (2002, p. 21) concluded, "A greater emphasis on understanding the epistemological differences between traditional and Western scientific knowledge systems should be explored in the middle years' (Grades 5 to 9) science programme, especially in schools with high Aboriginal populations." Such an understanding relates directly to the content of Chapters 3 to 7 in this book.

Conclusion

Science educators and Indigenous scholars worldwide are attempting to understand the ways by which marginalization in school science occurs for a student whose culture, language, and worldview differ from that embedded in Eurocentric school science. To alleviate the alienation experienced by many Indigenous students, cross-cultural science lessons and units developed by science educators encourage students to become knowledgeable, scientifically literate citizens and to pursue science-related occupations if they so choose.

Moreover, when teaching cross-cultural science lessons and units, instructors learn to shift their perspective of the two knowledge systems as mutually exclusive to one in which the knowledge systems are complementary (Chinn, 2007; Ogunniyi, 2007). Non-Indigenous teachers learn to build bridges between their own culture of Eurocentric science and their students' local Indigenous culture (Belczewski, 2009; Brayboy & Castagno, 2008; Cajete, 1999; Herbert, 2008). Indigenous science teachers who have not reflected much on their Eurocentric science education may go through a similar process (Chang & Rosiek, 2003).

The innovations we implemented when teaching science classes to Indigenous students over the years were highly successful with most non-Indigenous students as well. Teachers across Canada reported that response in non-science classes also improved when Indigenous knowledge became part of the science program and when their new-found expertise was transferred to those non-science classes (Coalition for the Advancement of Aboriginal Studies, 2002).

The under-representation of Indigenous students in science programs is not the sole issue. It is also about *the improvement in scientific literacy for both Indigenous and non-Indigenous students.*

CHAPTER 3

Eurocentric Science: Background

When people, such as Ms. Smith, the Grade 8 science teacher we discussed in Chapter 1, consider placing Indigenous knowledge in a science curriculum, serious discussion arises about whether Indigenous knowledge is a science or not. But what is a science? This chapter clarifies the term *science*, first by very briefly recounting its historical evolution into what it is today—*Eurocentric* science. This is followed by a summary of what most scientists do when they engage in Eurocentric science. We believe that Eurocentric science *is* what most scientists *do*. Scientific careers are extremely diverse and appear in a wide array of social-economic-political contexts. By describing a variety of these contexts, we are painting a realistic picture of the everyday involvement of Eurocentric science in society; we are not being critical of Eurocentric science.

In this book, scientists are recognized as practitioners of the scientific culture of their discipline. This culture is reflected in its expectations of its scientists. In turn, these expectations are conveyed by the following: (1) the technical language that precisely communicates their ideas, (2) the types of values that guide their professional work, (3) the standards that define achievement, (4) the beliefs they hold about reality, (5) the knowledge that characterizes their discipline, (6) the assumptions they make about that knowledge, and (7) the methods that produce their knowledge, including the technology they typically use. With an awareness of these cultural features of the scientific community, teachers like Ms. Smith will have a background for comparing Eurocentric science and Indigenous knowledge (see Chapter 7). However, before making the comparison, one must first have a realistic and accurate image of Eurocentric science. Stereotypes of Eurocentric science can prevent people from bridging the two cultures.

A Quick Look at the Historical Evolution of Eurocentric Science

The origins of science go back to ancient Egyptian and Greek philosophers. Greek ways of knowing nature were largely philosophical. This "pure" knowledge probably reached its zenith with Aristotle (about 330 BCE) or with Ptolemy (about 150 CE). Ideas and technology would later be taken from such places as China, India, and Islamic countries.

The evolution of Greek philosophy to modern science can be marked by major social changes in Europe. Understanding this clarifies what we understand or misunderstand science to be today. The first significant transformation was the Renaissance movement. In the twelfth century, Greek literature began to reach Europe via Islamic scholars in Spain, and this literature was translated into Latin. A few centuries later, these philosophical teachings about nature were taken up by people such as Galileo who called themselves *natural philosophers*. They began to assert a new authority of truth, one based on empirical evidence from observations and investigations, as opposed to one based on the authority of kings and the church (Mendelsohn, 1976). As time went on, this evidence-based knowledge system spread across Europe and Russia.

In the seventeenth century, natural philosophers such as Kepler, Descartes, Leibniz, Roberval, Huygens, Hooke, Halley, and Newton, firmly established this new knowledge system within European nations, based on empirical evidence and rational thinking. At that time, natural philosophy focused on the value of gaining power over nature. Natural philosophy became a social institution in England when the Royal Society, the first organization of natural philosophers, was founded in 1660. Other countries established similar organizations. That period in history is known today as the Scientific Revolution. European nations relied on the power of natural philosophy during their worldwide exploration and colonization efforts, which provided new frontiers for natural philosophy to explore. An example of this synergy can be seen in the voyage of Cook's ship *Endeavour* (1769–1770) in which natural philosophers were transported to Tahiti (Otahiti) to observe the transit of Venus across the path of the Sun. The voyage resulted in the "discovery" of the east coast of what is now known as Australia and led to the subsequent colonization of its Indigenous peoples.

The second social transformation in Europe arose from the success of natural philosophers at exercising control over nature. Entrepreneurs who adapted the methods of natural philosophy were able to increase control over human productivity during the establishment of various industries throughout eighteenth-century Britain (Mendelsohn, 1976). This adaptation gave rise to the Industrial Revolution, in which technology[10] became a powerful social force. Technologists regarded natural philosophy as the servant of technology. However, independent-minded natural philosophers disapproved. In the early nineteenth century, they began to distance themselves from technologists, thereby precipitating the next radical transformation in the evolution toward modern science (Mendelsohn, 1976).

Led by William Whewell, an Anglican priest and natural philosopher of mineralogy at Trinity College Cambridge, natural philosophers attempted to revise their public image by portraying technologists, such as James Watt of steam engine fame, as people whose success depended on the *application* of the abstract knowledge of natural philosophy. However, as in the case of the steam engine, the opposite actually happened. Technologist James Watt completed his invention long *before* scientists proposed the laws of thermodynamics that describe in abstract generalizations how the engine works. In other words, the technological success of the steam engine required a scientific explanation.

However, Whewell and his colleagues succeeded in their image-making. Today, there is widespread acceptance of technology as solely applied science (Constantinou, Hadjilouca, & Papadouris, 2010), thereby maintaining the ancient Greek philosophy that holds pure abstract ideas superior to practical knowledge (Collingridge, 1989).

Revising history was only one step in the nineteenth-century's radical advance toward modern science. A new social institution emerged to replace natural philosophy, and it sought to secure social importance in Britain. Some natural philosophers organized themselves into a *professional* institution (Orange, 1981). The word *science* was deliberately chosen to replace *natural philosophy* during the political birth of a new organization in 1831: the British Association for the Advancement of Science (BAAS). Since then, the term *science* has come to mean those scientific disciplines taught in universities. Today, it also means the field studies and research and

---

[10] The field of technology (in contrast to the field of natural philosophy or Eurocentric science) consists of people with disciplined capabilities who use technological knowledge (and occasionally scientific knowledge), materials, and human-social resources to respond to the needs of people and society. Prime examples today include engineers, some medical and agricultural practitioners, and technicians.

development (R&D) conducted by scientists employed in corporations, government agencies, the military, and private institutions. In archaic English, *science* simply meant "knowledge" (Latin: *scientia*). The BAAS added a new meaning of science to the English language, one which we primarily use today.[11]

The term *science* was politically chosen by the founders of the BAAS to set themselves apart from three different groups: natural philosophers, technologists steeped in the successes of the Industrial Revolution, and members of the old Royal Society (MacLeod & Collins, 1981). The BAAS also sought a politically privileged position from which to lobby financial support for the work of its members. It served as a model for the American Society of Geologists and Naturalists when, in 1848, the Society became the American Association for the Advancement of Science (AAAS), the major scientific organization in the United States today.

In 1867, the BAAS helped establish the first known English school science curriculum. The organization ensured that the curriculum's ideology supported the following:

- an elite upper class of students
- pre-professional screening for university science departments
- an emphasis on mental training and abstract knowledge over practical know-how.

As a result of the evolution of natural philosophy into professional science, present-day science is strongly based on Euro-American thinking. Scientists collectively work within a culture that frames their thinking and practice in the context of their professional work. As described in the section "The Culture of Eurocentric Science" later in this chapter, most scientists' professional culture is Eurocentric in character, and can be described as *Eurocentric science* or *Western science*.

Scientists: Who Are They?

Eurocentric science today is conducted by a *community* of people who contribute collaboratively to its evolution and to its knowledge system. The

---

[11] The term *scientist* was coined by Whewell in 1834 to describe members of BAAS. But, they insisted on calling themselves *men of science* well into the twentieth century (MacLeod & Collins, 1981).

days of the lone scientist/natural philosopher—heroes of the past like Newton—are over.

Academic science, sometimes referred to as "pure" science—research without obvious commercial benefit—consists of teams of scientists worldwide who closely keep in touch with one another's work in a cooperative-competitive dance. They communicate openly but are cautious about revealing too much information in case other research teams solve a puzzle with it and publish their results or apply for funding first. These informal clusters of scientific teams are often called *invisible colleges*.

At conferences, members of an invisible college do not simply exchange facts. According to Nobel laureate John Polanyi (2009), scientists also

> ... advance propositions. As with evidence in court, these are tested in cross-examination before a jury of our peers. Truths established in this fashion can subsequently be overthrown by a higher court. Indeed, on examination, the new laws of science often turn out to be approximations. (p. A15)

In short, established "truths" arise among teams of scientists rigorously working toward a *consensus* of what to believe at the present time, a process described in Chapter 4 (the sections "Diversity within Eurocentric Sciences," and "A More Realistic View of Eurocentric Sciences"). Scientists are skeptical people who scrutinize another research team's methods, empirical data, and the interpretation of those data. Then they reach a consensus. Consensus does not mean unanimous agreement.

Today, most scientists are trained by academic science departments at universities throughout the world and are subsequently employed by such institutions as corporations, government agencies, health centres, and private foundations. There, they continue to learn occupation-appropriate knowledge and skills. Many work on R&D projects, in which scientists are paid by their institutions to carry out investigations to develop and apply knowledge that benefits those institutions. Although R&D is much different from academic science, scientists engaged in such work maintain the same high standards in their investigations.

In his book *Real Science: What It Is and What It Means*, renowned physicist John Ziman (2000) discusses "post-academic" science. Since about 1950, post-academic science has by and large replaced academic "pure" science in terms of the production of new knowledge. "Increasing globalization, together with an awareness of the need for sustainability, accelerated the merging of industrial and academic science into post-academic science ... largely driven by industrial and governmental needs" (Levinson, 2010,

p. 77). Post-academic science is both transdisciplinary and utilitarian, producing value for the money spent on it. The choice of subjects scientists investigate is strongly influenced by policies formulated by corporations and politicians. Post-academic science may be said to be both industrialized and bureaucratic (Wong & Hodson, 2010; Ziman, 2000) and is the essence of R&D. Thus, post-academic science represents *a more realistic description of Eurocentric science* than does academic science.

Only a small proportion of scientists finds employment in universities (Ziman, 1984, 2000). These academic scientists usually have more freedom to decide what puzzles to investigate. Some use their expertise to collaborate with business and industry because that is where their research money is found (Bencze, 2008). Some choose to follow their unfettered curiosity like Steven Hawking in his search for a theory of the origin of the universe. Periodically, they make far-reaching, practical contributions. For example, some academic scientists surmised that the melting of a chocolate bar in a colleague's shirt pocket was caused by energy emitted from a new instrument that identified flying aircraft (a World War II military project). Their investigations eventually led to the production of microwave ovens. Some academic scientists respond to a specific societal need as typified by chemistry professor Lee Wilson (of Métis ancestry) who investigates nano-sized materials that filter water to get rid of contaminants (Hounjet, Kvamme, Mohr, Phillipchuk, & View, 2011). His motivation stems from the serious illness of his father caused by drinking contaminated water. Dr. Wilson's research is designed to help isolated communities maintain the quality of their drinking water. As a scientist, he bridges his Métis culture and the culture of chemistry. Other academic scientists collaborate specifically with Indigenous groups; examples include scientists in a field called traditional ecological knowledge and wisdom (TEKW) (Menzies, 2006), the environmental audit model developed by the Walpole First Nation, and a Renewable Energy Certificate program developed by the Seven Generations Education Institute (2008)—a collective of Anishinaabe Nations in Ontario.

Whether engaged with corporate, government, or academic science, *all* scientists are part of specific communities of practice. These include their immediate colleagues at work, their invisible college, and their professional scientific organizations such as the Canadian Society for Chemistry, and the American Association for the Advancement of Science (AAAS). Professional scientific organizations exist to support professional activities, such as helping with job placement, reviewing manuscripts, publishing journals, and organizing conferences, and to lobby federal governments on behalf of members.

Like the image of Newton as the lone scientist hero, the days of "pure" science no longer exist (Wong & Hodson, 2010). In *Never Pure*, Harvard professor of history of science Stephen Shapin (2010) goes further and argues that science has never been pure. Today, as much as 95% of funded scientific research is linked to specific societal outcomes (Ziman, 2000). Major advances in biotechnology, nanotechnology, molecular biology, materials science, and light source synchrotron investigations, to name just a few, generate new knowledge and support the development of new products. These advances create wealth for transnational corporations in the global economy. Globalization and Eurocentric science are intricately linked in the production of wealth.

Scientists: Who are they? A few are academic scientists, while most are post-academic scientists conducting R&D projects for commercial, government, and health-oriented institutions. Great diversity exists within the employment of scientists.

Other Views of Eurocentric Science

Consider the view held by many Indigenous peoples worldwide regarding Eurocentric science. In past centuries, Indigenous nations experienced the powerful influence of Eurocentric science through ships, cannons, and guns used for exploration, oppression, and colonization. Descendants of these Indigenous people regard Eurocentric science and colonization as interconnected (Davis, 2009; Sillitoe, 2007). For contemporary Indigenous peoples, Eurocentric science and technology offer such advantages as electricity, GPS technology, immunization, and improved transportation and communications. But, these are offset by negative aspects, such as agribusiness, resource extraction, and biopiracy (Settee, 2000). Viewing the Hollywood movie *Avatar* with an Elder and discussing it afterwards would be instructive. Many First Nations, Métis, and Inuit Elders talk about Eurocentric science in terms of commercial commodities (Elliott, 2008).

The general public usually perceives scientists as people who fulfill a utilitarian function by inventing things for the benefit of society (Ryan & Aikenhead, 1992). Scientists are seen less as people who have an explanatory function that describes, explains, and predicts natural phenomena.

Of course, scientists usually serve both functions, depending on the institution they work for. Advances in Eurocentric science have occurred for many different reasons but primarily around the needs and nature of society.

The explanatory function of Eurocentric science is of special interest to science teachers since explanation is heavily emphasized in conventional school science curricula. As an alternative, science-technology-society-environment (STSE) curricula include the utilitarian function to some degree with their emphasis on Eurocentric science embedded in society and how decisions about science and technological projects must include social and cultural dimensions.

The idea that Eurocentric science serves *only* a utilitarian function is more applicable to either the field of technology or the combination of Eurocentric science and technology known as R&D. This utilitarian bias can be found in the media. For instance, a newspaper story concerning a recent press release from a scientific institution translated "technological achievements" (e.g., the steam engine) by using misleading terms such as "scientific inventions" or "scientific breakthroughs."

The public's emphasis on utilitarian function differs from the curriculum's explanatory emphasis, but either position is correct, depending on the context in which Eurocentric science is practised. What is not justified, however, is the myth that technology is *simply* the application of science, a misconception discussed in the section "A Quick Look at the Historical Evolution of Eurocentric Science." In R&D, Eurocentric science is as much applied technology as technology is applied science (Bencze, 2008; Ziman, 1984). Eurocentric science and technology are mutually enriching.

Approaching Eurocentric science from a cultural point of view helps science teachers bridge cultures for the purpose of including some Indigenous knowledge in their science instruction (Belczewski, 2009). The key to understanding this cultural topic is the concept of worldview.

Worldview

To the layperson, *worldview* is the way people experience and make sense of the world. It is the way they see, interpret, understand, experience, and react to what is around them. Worldviews interest science teachers because teachers are constantly trying to engage students whose worldviews often differ profoundly from the scientific worldview conveyed in the classroom.

The anthropological meaning of worldview is the one most commonly found in science education as "a set of assumptions and beliefs that form the basis of a people's comprehension of the world" (Cajete, 2000b, p. 62). A compatible but more detailed definition comes from Cobern (2000): "A worldview refers to the culturally dependent, *implicit*, fundamental organization of the mind. This implicit organization is composed of presuppositions that predispose one to feel, think, and act in predictable patterns" (p. 8). For example, a science teacher who appreciates a student's worldview will likely anticipate which ideas in a science curriculum may appear plausible to the student, and which ones may not. As Cobern (1996) also mentions:

> Worldview provides a nonrational foundation for thought, emotion, and behaviour. Worldview provides a person with presuppositions about what the world is really like and what constitutes valid and important knowledge about the world. (p. 584)

These meanings of worldview all apply to our discussions of Eurocentric science and Indigenous knowledge. A comparison of each group's collective worldview was compiled by Barnhardt (2006). The Canadian Council on Learning (2007a) published three "holistic lifelong learning models," one each for First Nations, Inuit, and Métis peoples, that express each of their collective worldviews. A worldview diagram of the Sweetgrass Cree First Nation was developed by Traditional Knowledge Keeper Judy Bear (Pearson Canada Inc., 2010, p. PO-67). McGilchrist (2009) gives a neuroscience account of generalized collective worldviews in terms of the brain's two hemispheres.

The expressions *the scientific worldview* and *the Indigenous worldview* are misleading because they create a stereotype of each group. They ignore the variations of worldviews within the scientific community (see Chapter 4) and among Indigenous peoples (see Chapters 5 and 6). The expressions *a scientific worldview* and *an Indigenous worldview* are more accurate, but might convey a stereotypical image as well. More precise descriptions, such as *a scientifically compatible worldview* or *a worldview endemic to Eurocentric science*, and *a worldview held or recognized by an Indigenous group* or *a worldview compatible with Indigenous peoples* could be used. But these expressions are wordy and cumbersome. The expressions *a scientific worldview* and *an indigenous worldview* are intended to convey a non-stereotypic meaning (as the more wordy expressions mentioned above suggest).

The Culture of Eurocentric Science

As eminent chemist Carl Djerassi (1998, p. 511) explains, "Scientists operate within a tribal culture ... acquired through intellectual osmosis in a mentor-disciple relationship." A scientific culture depends on expectations shared by its practitioners. These expectations are like a series of cultural performances that express certain values and demonstrate expertise (Shapin, 2010). In an experimental investigation, for example, variables are rigorously controlled and parameters of the study are clearly defined. Except when national security and corporate secrecy are relevant, results are presented publicly in a form that often permits mathematical analysis. Experimental studies should be replicable by other investigators at different sites. For any non-confidential investigation, methods and results are typically reported for evaluation through peer review (described in Chapter 4, the section "Diversity within Eurocentric Science").

Eurocentric science is known in some fields of cultural anthropology as the "culture of Western science" (Pickering, 1992; Sillitoe, 2007), albeit a different type of culture from that conventionally associated with ethnicity. As described earlier in this chapter, a cultural perspective on Eurocentric science emphasizes its norms, values, beliefs, knowledge, language, technologies, expectations, and conventional actions as shared by a community of scientists. Eurocentric science is inhabited by many communities of practice "enmeshed in a network of cultural communication" (Brandt, 2008, p. 842). The foundation of Eurocentric science is Euro-American culture and it conveys a Eurocentric worldview (see the section "A Quick Look at the Historical Evolution of Eurocentric Science"), and described in detail throughout Chapter 4. The culture of Eurocentric science certainly has a very powerful way of knowing nature.

Eurocentric science includes knowledge accumulated over the ages from many non-European cultures (e.g., Eastern and Islamic cultures). However, knowledge from these other cultures has been modified to better fit the worldview, metaphysics, ways of knowing, and value systems in the culture of Eurocentric science (Harding, 1998).

Two specific implications for science teachers follow from this fact. The first centres on the term *ethnoscience*. For example, Indigenous knowledge about the healing power of plants is often called *ethnobotany*. This label creates a dichotomy between botany and ethnobotany, which empowers botany over the special case of ethnobotany. But, as we have just seen, botany taught at a university has an ethnicity—Euro-American. It, too, is an ethnoscience. Therefore, the word *ethnoscience* applies to all knowledge systems that explain nature in culture-based terms. Consequently, the dichotomy of botany versus ethnobotany is a false dichotomy (described in Chapter 1 as something best avoided). Such words as *ethnoscience* should be dropped from our vocabulary. Labels have extraordinary power and may give prominence to one viewpoint at the expense of another.

The second implication raises the question of who can become a scientist. Although people from any non-Euro-American ethnic group worldwide can become scientists, their university degrees prepare them to act in accordance with the culture of Eurocentric science. Even successful Indigenous scientists have experienced this process of enculturation into Eurocentric science—the mentor-disciple relationship Djerassi (1998) wrote about. Science is universal in that anyone may apply, but everyone is expected to conform to certain rules: the group's shared norms, values, beliefs, canonical knowledge, language, technologies, expectations, and conventional actions. If a person is unwilling or incapable of conforming, that person risks not being hired, funded, or published. In short, Eurocentric science *appears to be* universal but comes with scientific cultural strings attached.

The Problem with Universalism

In the context of science education, universalism is a doctrine comprising several different assumptions, which are examined in Chapter 4. Universalism appears in university undergraduate science courses, the media, and many classroom resources. Unfortunately, it has a negative impact on teaching Indigenous knowledge in school science (see Chapter 4, the sections "Reality Is Reproduced or Represented by Scientific Knowledge," "Empirical Data Speak for Themselves: Positivism," and "Conclusion").

In addition to assuming that any qualified person can become a scientist, universalism holds that when scientists observe reality, the result is a true image of reality or else a very close facsimile, not a *representation* of reality mediated by human perception, imagination, social conventions, and cultural predispositions. In other words, the scientific result is believed to be universal in the sense that human elements do not interfere; therefore, only one true *universal* result is possible—the one dictated by reality or scientific evidence. Thus, universalism holds that scientific knowledge transcends culture because it has no cultural content—it is culture free. Furthermore, universalism assumes that reality does not change over time or in different places in the universe. Thus, scientific knowledge, such as chemical equations in molecular biology, is generalizable and can be applied universally and largely without concern for context (Chapter 4, the section "Scientific Knowledge Is Generalizable"). Generalizability is one of the many strengths of Eurocentric science, but has limitations when compared with Indigenous local knowledge in certain circumstances, as seen later in Chapter 6 (the section "Valid").

Defenders of the doctrine of universalism point to the ability of Eurocentric science to explain, predict, and control natural phenomena. Universalists conclude that they have the most effective, comprehensive, and reliable way to know nature. Any other way, such as Indigenous knowledge, is inferior and should be ignored.[12] The negative consequences of the combination of universalism and globalization, known as "global science," are documented thoroughly by Sillitoe (2007). Chapter 4 continues this discussion and concludes that universalism is only one of several cultural doctrines embraced by some, but certainly not all, scientists.

An alternative to universalism is the idea that there are multiple legitimate ways of understanding nature as seen in the clarification of the word *science*.

Clarification of *Science*

Although labels can create privilege in one group over another, they also have the power to create equity. The label *Eurocentric science* automatically opens our mind to types of science that are not Eurocentric, and leads us to explore those alternatives. But first we must clarify what the generic term *science* can mean. By doing so, we continue to focus on the science curriculum's explanatory function, as well as on a teachers' goal to build cross-cultural bridges so that students learn the best of both cultural ways of knowing nature.

Japanese science educator Masakata Ogawa (1995) thought of a generic meaning for science: *a rational, culturally based, empirically sound way of knowing nature that yields, in part, descriptions and explanations of nature.* In other words, all cultures worldwide have a science. This encompassing generic concept of science includes several perspectives, such as the following:

---

[12] See Stanley and Brickhouse (1994, 2001), Gough (2002), or McKinley (2007) for a thorough discussion on universalism.

- a Euro-American culture-based perspective (Eurocentric science)
- various Indigenous ways of knowing nature, including Indigenous knowledge, "Mäori science" (McKinley, 1996), "Yupiaq science" (Kawagley, 1995), "African science" (Keane, 2008), and "West Indies science" (George, 1999; Herbert, 2008)
- non-Euro-American perspectives known as "neo-indigenous ways of knowing nature" (Aikenhead & Ogawa, 2007), which include Islamic science (Irzik, 1998; Loo, 2001; Sadar, 1997), Chinese science (Needham, 1956), and a Japanese way of knowing nature (Aikenhead & Ogawa, 2007). These neo-indigenous perspectives will interest urban teachers in multicultural classrooms.

In science education, our definition of *science* encompasses several empirically based ways of knowing nature. Such a definition gives us an equitable perspective on ways of knowing nature, a perspective that Mäori science educator Elizabeth McKinley (2007) calls "pluralism." Those committed to universalism call it *relativism*—the notion that any idea is acceptable—which is not the meaning of pluralism. Scientists and their related professional organizations may likely not adopt this pluralist definition of science because their professional identities typically embrace a singular meaning of science. The public will likely not adopt a pluralist definition either. Our purpose here is not to change how scientists and the public define science; it is to help teachers bridge two cultures to assist students in school science.

We assume a pluralist meaning of science because it fosters the teaching of school science in culturally responsive ways (see Chapter 8). This pluralist perspective corresponds to the common-sense way Grade 7 northern Manitoba Cree students understand the word *science* (Sutherland & Dennick, 2002). Similarly, the Native American scholar Greg Cajete (2006, p. 248) defines science as "a story of the world and a practiced way of living it." His expression "Native science" is his way of naming Indigenous knowledge to create equity between it and Eurocentric science, and his way of highlighting his peoples' "creative participation with the natural world in both thought and practice" (p. 249).

Earlier in this chapter, MacLeod and Collins's (1981) historical account of the formation of the BAAS placed the word *science* squarely in a political arena of elite social privilege. This book revisits that arena but in the context of twenty-first-century science education. Canadian and Australian prime ministers have apologized to Indigenous peoples for the oppressive treatment many suffered in residential schools and its lingering consequences today. The prime ministers ushered in a new era of reconciliation and mutual respect.

Accordingly, we broaden the 1831 meaning of science by adopting Ogawa's (1995) pluralist perspective. This shift from universalism to pluralism is a crucial step for teachers to take (Gaskell, 2003). It is the difference between a science teacher who rejects the inclusion of Indigenous knowledge in the science curriculum and one who is open to exploring its inclusion. The concept of pluralism itself forms a major bridge between Eurocentric science and Indigenous knowledge.

A pluralist meaning of science implies that any time the word *science* is used in this book, we must qualify what we mean. We do this by placing an adjective in front of the word *science*: Eurocentric science, school science, and so on.

The phrase *science education* signifies a pluralist perspective, which encompasses both Eurocentric science and Indigenous knowledge. For teachers of multicultural classrooms, neo-indigenous knowledge may be included as well. Conversely, a perspective adhering to universalism is known as *conventional science education*, because the media, many classroom resources, and university science departments conventionally express a universalist viewpoint (see Chapter 7, the section "Deborah's Story," and Chapter 8, "Deborah's Story Revisited").

In summary, this chapter recognizes the diversity within scientific employment; it describes some strengths and limitations of Eurocentric science, and clarifies the pluralism of sciences worldwide. This book offers a view of Eurocentric science that acknowledges the conventional academic view, but it also recognizes the realities of professional scientists and their scientific enterprise in the everyday world. Some of these realities were originally identified by Indigenous writers. We have recognized realities not as a critique of Eurocentric science, but as a clarification of a more encompassing and realistic view, which will help teachers bridge the cultures of Eurocentric science and Indigenous students. If we attempted to connect only the conventional academic view of Eurocentric science with the realities of Indigenous cultural ways of knowing nature, our message would not be credible.

Chapter 4 provides further details about what Eurocentric science is and is not, in short, realities and myths.

CHAPTER 4

Eurocentric Sciences

How diverse is Eurocentric science? Do scientists really use one scientific method? Can Eurocentric science actually be objective? What presuppositions guide the way scientists observe and think? Answers to these and other questions further define what Eurocentric science is and lead science teachers to a clear comparison between it and Indigenous knowledge (Chapter 7).

Diversity within Eurocentric Science

One of the most well-known descriptions of Eurocentric science is Thomas Kuhn's *The Structure of Scientific Revolutions* written in 1962. In this classic book, Kuhn expands our understanding of Eurocentric science by including historical and sociological dimensions in academic scientists' professional practice. His ideas, discussed below, helped identify the subjective human elements in scientific ways of knowing and in the fabric of scientific knowledge itself.

A scientific "paradigm," or a framework for thought and action, determines what a community of like-minded scientists thinks and how they conduct research. One example of such a community would be a group of scientists who explore ideas about the origins of life on the planet by conducting biochemical laboratory experiments. A different paradigm might act as the framework for another group of origin-of-life scientists conducting DNA investigations (Hazen, 2005). The invisible colleges described in Chapter 3 usually share a paradigm.

Groups of scientists working within a paradigm belong to a subculture of Eurocentric science. Kuhn's ideas help to characterize the great diversity found within these subcultures. For example, many scientists engage in what Kuhn calls "normal science"—puzzle solving through the use of established science content and practices, also known as paradigm-directed research. Other scientists may engage in "extraordinary science" or paradigm-shattering research that could lead to challenging scientists' allegiances to a paradigm. Moreover, paradigms can be so diverse that communication is hampered between different groups of scientists, such as one of deep ecology researchers[13] and another of biochemical medical researchers. Each group may have different accepted meanings of words and ideas within each paradigm. Kuhn called these kinds of paradigms "incommensurate."

In the second edition of Kuhn's book published in 1970, he acknowledged that his term "paradigm" had several legitimate meanings. To clarify this ambiguity, he added a postscript with four major points.

First, he reiterated that a paradigm is associated with a group of like-minded scientists who produce and validate scientific knowledge. "A paradigm governs, in the first instance, not a subject matter but rather a group of practitioners. Any study of paradigm-directed or paradigm-shattering research must begin by locating the responsible group or groups" (p. 180). Validation of ideas within a paradigm requires rigorous data evaluation, argument, and consensus making among the group that subscribes to the paradigm. This includes the scientists who review scientific manuscripts to determine whether a manuscript will be published (see the section "A More Realistic View of Eurocentric Sciences"). This peer-review process decides which ideas will generally be accepted as "true" for the time being.

Second, Kuhn replaced the term "paradigm" with "disciplinary matrix" to clarify a further meaning of the term. A disciplinary matrix is a "constellation of group commitments" (p. 181). This constellation comprises the following:

- symbolic generalizations (e.g., vocabulary and equations);
- beliefs in particular models of reality;
- values that guide scientists when they judge evidence and theories, as well as values that scientists draw upon to reach a consensus on what to believe as "true;" and
- problem-solving exemplars.

---

[13] Deep ecology is a radical field within environmental studies that considers humanity as an organic whole within "the Earth household," in which "the spiritual and the material aspects of reality fuse together" (Devall & Sessions, 1999, p. 200).

A third meaning of paradigm is an "assimilated, ... time-tested, and group-licensed way of seeing" (p. 189) natural phenomena. This tacit knowledge is acquired only by doing Eurocentric science, especially under the guidance of "a mentor-disciple relationship" as Djerassi (1998, p. 511) described it.

This leads to Kuhn's fourth point: Scientists' observations and interpretations depend upon scientists' prior experiences and training (p. 198). Scientists using empirical data are influenced by their own prior experiences, training, and the theories they hold. In short, scientific observations are *theory-laden* (Goldstein & Goldstein, 1981). Kuhn described situations where the same natural event was observed differently by scientists in two different paradigms that influenced what was observed. The observations were 'paradigm-laden.' In such situations, scientists of one paradigm can have difficulty communicating with scientists from another paradigm, resulting in misunderstandings, that is, incommensurate paradigms.

Kuhn demonstrated that Eurocentric science does not proceed in a purely logical and impersonal way. His detractors, however, accused him of undermining the authority of Eurocentric science by placing it at the mercy of human emotions and intellectual fads. This criticism was countered by Bauer (1992) who pointed out that scientific consensus making emphatically relies on critically analyzed empirical data, and, therefore, consensus making is based on more than subjective group commitments to a paradigm. But he also pointed out that the recent history of Eurocentric science "offered ample instances where science *did* incorporate false beliefs, sometimes under the influence of emotion and fashion" (p. 62).

The existence of multiple paradigms, some of which may be incommensurate, illustrates the extensive diversity within the scientific enterprise. A more authentic term to label this diverse culture-based scientific enterprise may be appropriate. The plural phrase *Eurocentric sciences*, abbreviated as ES, will be used throughout the remainder of this book.

Paradigms can be diverse even within a single scientific field. For example, consider the origin-of-life research described by Hazen (2005) in his book *Genesis: The Scientific Quest for Life's Origin*:

> Scientists crave an unambiguous definition of life, and they adopt two complementary approaches in their efforts to distinguish that which is alive from that which is not. Many scientists adopt the "top-down" approach. They scrutinize all manner of unambiguous living and fossil organisms to identify the most primitive entities that are, or were, alive. For origin-of-life researchers,

primitive microbes and ancient microfossils have the potential to provide relevant clues about life's early chemistry.... (p. 26)

By contrast, a small army of investigators pursues the so-called "bottom-up" approach. They devise laboratory experiments to mimic the emergent chemistry of ancient Earth environments. Eventually, the bottom-up goal is to create a living chemical system in the laboratory from scratch—an effort that might clarify the transition from nonlife to life. Such research leads to an amusing range of passionate opinions regarding what is alive, because each scientist tends to define life in terms of his or her own chosen specialty [paradigm]. (p. 27)

The "amusing range of passionate opinions" refers to caustic public debates between some scientists that Hazen chronicles in his book. These scientists demonstrate passionate commitments to different paradigms within the origin-of-life field of research.

ES today are strengthened by the diversity of paradigms within a particular scientific field and the background diversity of scientists working as a team. Collaboration between corporate and academic scientists illustrates this strength. For example, the world-class Canadian Light Source synchrotron was built in Saskatoon under such a collaboration. Given the paradigm diversities within ES, the scientific enterprise is not uniformly straightforward.

The Myth of *the* Scientific Method

Different paradigms demand different research methods. These vary so widely that it would be unwise to think that a single, logical, five-step method—*the* scientific method—could represent all Eurocentric sciences. This notion of a lock-step scientific method appears in many science classroom materials[14] and is consequently widely shared by the general public. According to Rudolph (2005), the phrase "*the* scientific method" was misappropriated years ago by science educators who read John Dewey's 1910 book *How We Think*. The ubiquitous existence of "the scientific method" in today's schools, universities, and media suggests that this idea continues to pervade people's thinking about ES and has become a stereotype, which conveys an idealized view. It evokes the caricature of a bespectacled old

---

[14] Between five and seven steps are often identified, depending on the source. The following is a typical example: (1) state the problem, (2) state a hypothesis as to the cause of the problem, (3) design an experiment to test the hypothesis, (4) predict the results of the experiment, (5) conduct the experiment to observe the results, and (6) make conclusions based on those results.

man in a white lab coat. The concept of the scientific method has acquired the status of an urban myth. However, this myth is not supported by active scientists (Wong and Hodson, 2009). Many scientists and scholars, including Dewey himself, have denounced it. A more appropriate perspective is described in detail in the section to follow ("A More Realistic View of Eurocentric Sciences").

Bauer in his book, *Scientific Literacy and the Myth of the Scientific Method* (1992), identifies variations in different sorts of ES: young/mature, data-driven/theory-driven, data-rich/data-poor, experimental/observational, and quantitative/qualitative. Any one of these variations can influence the scientific methods used by a team of scientists.

Research in science classrooms (Tang, Coffey, Elby, & Levin, 2010) shows that using a lock-step scientific method has a *detrimental* effect on students' achievement in scientific inquiry. Students may pursue scientific inquiry very successfully without the distraction of "the scientific method." An alternate guide for classroom investigations is "model-based inquiry," which emphasizes classroom activities and discussions of content that is testable, revisable, explanatory, conjectural, and generates further investigations (Windschitl, Thompson, & Braaten, 2008). Instead of "the scientific method," science educators could rely on an authentic set of *values* shared among most scientific paradigms. Such an approach includes skepticism and a demand for evidence, precision, and a critical analysis of assumptions.

A More Realistic View of Eurocentric Sciences

How do Eurocentric sciences (ES) really work? Bauer (1992, pp. 43–44) points out some key features, including the following:

- Modern science began with widespread and systematic cooperation among scientists (i.e., natural philosophers) during the Scientific Revolution.
- What ultimately distinguishes pseudoscience from Eurocentric sciences is not the methodology or subject matter, but pseudoscience's *isolation from the scientific community*.

Along with Kuhn, Bauer describes legitimate ES operating within a community of scientists, rather than in isolation. Attempts by philosophers to demarcate pseudoscience from legitimate Eurocentric sciences have failed (Gieryn, 1999). This book defines Eurocentric sciences as *what scientists do*. Hence, ES are communal and exist within groups of people (Shapin, 2010; Ziman, 1984). In science classrooms, this understanding is essential if students are going to critically analyze science-based news items reported in the media (McClune & Jarman, 2010).

What scientists do depends, in part, on their paradigm, invisible college, or research team. *Theoretical* scientists like Stephen Hawking pose mind-boggling puzzles, such as how did the universe begin, and they use mathematical methods, ingenuity, and complex rational reasoning to construct a "theory of everything" (Greene, 1999, p. 366). Some theoretical scientists even develop computer models that generate their own data.

Most scientists, however, fall into the category of *empirical* scientists. Their investigative methods gather data through experiments, field studies (e.g., geological, ecological, and resource management), and quality-control monitoring procedures typically used by industrial, government, and medical scientists. Their methods, results, and interpretations are reported. If necessary and when possible, empirical scientists repeat other teams' investigations to ensure the same results can be replicated in their own facility. Interdisciplinary studies are common today and require a research team of scientists from different fields. Scientists within the same team may use very different methods depending upon what part of the project they work on.

Experimental scientists, in particular, typically proceed in ways that include, but are not restricted to, the following strategies:

- Become curious about something or be given a puzzle to solve.
- Search out information (review the literature) and any other ideas about the puzzle.
- Propose a hypothesis to guide the design of experiments that may lead to solving the puzzle.
- Decide what variables might be relevant, and design experiments to control some variables and measure others.
- Acquire funding to pay for personnel, equipment, etc.
- Make sure all instruments work properly.
- Carry out carefully controlled experiments to determine if the hypothesis accurately predicted the results (and repeat this strategy as needed),
- Write up a research report for the funding agency, and another for public dissemination (unless precluded by corporate or government confidentiality) in peer-reviewed journals.

To write up their investigations, scientists rely on empirical evidence, quantitative or qualitative, and on the conclusions drawn from that evidence to formulate a rigorous argument.

The resulting argument must answer a key question: *Should the scientific idea (conclusion) be accepted?* Such an argument must survive the scrutiny of skeptical colleagues demanding precision, consistency, trustworthy evidence, logic, clear and correct terminology, along with a critical analysis of assumptions. In short, the writers need to persuade a reader who will use these same stringent values to assess the report. This act of persuasion reduces the subjectivity that a team might inadvertently bring to writing their report. As described by 254 eminent scientists in the U.S. National Academy of Sciences (Gleick, 2010), the process is adversarial; scientists' reputations are at stake; and credibility can depend on successfully challenging the status quo or on demonstrating the validity of a hypothesis.

Writing clear and convincing arguments is an important skill for scientists to acquire and one that should be encouraged in school science by teaching students how to write a scientific argument (McNeill, 2009).

A team's research report may eventually reach a formal consensus-making process, perhaps privately among scientists in a corporate R&D department or perhaps publicly among scientists on a peer-review panel determining whether the report will be published. A scientific journal's peer-review panel will often analyze the report by asking: Is the evidence based on reliable and accurate data? Is the new scientific idea or conclusion consistent with other scientific ideas, or are there serious discrepancies? Does the idea make a novel contribution to the scientific community? Is the idea simple, not unnecessarily complicated, and widely applicable? Does the idea suggest new types of investigations not thought of before? The answers to such a list of questions form a repertoire of values that guide most empirical scientists.

A realistic view of ES leads to this important conclusion: Rather than sharing common scientific methods, ES generally share common scientific *values*. ES are 'value-laden' (Goldstein & Goldstein, 1981; Shapin, 2010; Ziman, 1984).

Another key question can arise: *Should the scientific conclusion be acted upon?* Empirical investigations may be quality-control studies, medical studies, and environmental impact studies. These have social, economic, political, legal, or ethical ramifications. The answer to this key question necessitates a set of strategies for gaining and evaluating information. The strategies are outlined in a "model for measurement, data, and evidence" (Gott, Duggan, & Roberts, 2007) and can be illustrated by a simple medical example. Suppose a nurse in a surgical ward takes a patient's temperature (Level 1 in the model: *measurement*). Because the temperature is high, the nurse repeats the measurement with a different thermometer. With confidence in this high temperature measurement, the nurse then "takes it to the next level" (Aikenhead, 2005) by measuring other variables such as blood oxygenation and blood pressure. The collection of these diverse measurements is Level 2: *data*. Now the nurse, using his or her past experiences and education, can look for a pattern in the data and possibly relate it to information (data) about the patient's post-surgery condition. With these added data, credibility in medical reasoning increases, and the nurse is at Level 3: *evidence*. The nurse may believe that the evidence is strong enough to warrant a special test for the patient. Alternatively, the nurse may no longer be concerned about the high temperature but will continue to monitor the patient. Suppose the evidence was acted upon by getting a special test, but the results were ambiguous. The situation then approaches Level 4: *wider societal issues*. Does the nurse "take it to the next level" where there are social-ethical ramifications? The nurse must decide whether the evidence is credible enough to require the intervention of a medical doctor. The particular social issue in hospitals focuses on the limited number of available doctors and the consequences to the nurse if a doctor is unnecessarily called in to decide what action to take. On the other hand, the ethical issue highlights the probability of the patient's condition worsening. It is at Level 4 that the nurse consciously combines scientific thinking about measurement, data, and evidence with, in this case, social-ethical thinking.

This four-level model (measurement, data, evidence, and wider societal issues) was derived from empirical research in a variety of industries, and undertaken with scientists working within the social context of their employer (Gott, Duggan, Roberts, & Hussain, 2009). As a result, the model fits many scientific situations, and is applicable to most employees working in science-related occupations. It promotes learning 'concepts of evidence' used to critically evaluate scientific data and evidence—concepts such as validity, reliability, causation, correlation, probability, and risk. The model offers a realistic view of Eurocentric sciences in action. It could help students analyze more complex issues such as climate change research.

There are implications for science teachers and curriculum developers who believe in conveying a realistic and relevant view of ES to students.

Classroom activities and assignments, such as guided inquiry, lab reports, scientific case studies, and research papers, should

- articulate scientific values,
- embed scientific practice in local or global social-economic-political-legal-ethical contexts, and
- focus on teaching concepts of evidence (Aikenhead, 2006).

This type of school science relates to *what scientists do* in the everyday world of post-academic science, and it illustrates our definition of ES. Examples can be found in approaches to teaching known as science-technology-society-environment (STSE), the nature of science (NOS), and socio-scientific issues (SSI).

The Myth of Achieving Objectivity

As described in the previous section, peer review and consensus making within a community of scientists determine what "scientific truth" is at any one time. During this process, scientists scrutinize ideas, methods, data, and arguments. Eventually, they reach a conclusion, but not necessarily a unanimous one. Consensus making moves scientists *toward* the ideal of objectivity, although they can never achieve it (Wong & Hodson, 2009; Lewontin, 1991; Ziman, 1984). Consensus making *reduces* the subjectivity of individual scientists or teams of scientists. Consequently, a realistic goal for scientists is *low subjectivity*. The public storyline that scientists attain objectivity is a myth (Goldstein & Goldstein, 1981). The ideal of objectivity fails in the reality of practice (Lewontin, 1991). Nevertheless, objectivity remains a powerful and useful ideal.

In Holton's (1978) book, *The Scientific Imagination*, he includes several case studies to show how such subjectivity as intuition, imagination, and personal–social values or idiosyncrasies of scientists can propel ES forward and alter their direction. When subjectivity leads to exciting new avenues for investigation and alternative methods of analysis, there is always the scrutiny of consensus making and its rigorous scientific values as described earlier. This was summed up beautifully by a scientist quoted in the study of Wong and Hodson (2009, p.119): "Assume with daring; verify with care." Active scientists agree that creativity and imagination are important at all stages of the scientific process.

Some scientists work at the frontier of knowledge, called *frontier science* (Cole, 1992), where the evidence may be weak and the knowledge preliminary. For example, scientific knowledge about the link between electromagnetic waves associated with cellphone use and cancer is so incomplete that current conclusions appear to be conjectures (Albe, 2008). However, scientific knowledge about electromagnetic waves is unlikely to change much in the foreseeable future. This type of knowledge is associated with *core science* (Cole, 1992). In general, the type of puzzle scientists are working on in either frontier or core science will determine the degree of subjectivity involved.

When there is less certainty in the knowledge as in frontier science, controversy often arises along with a greater need to determine the credibility of the sources of scientific information. As a result, there is a greater reliance on values related to "Who can you trust?" Frontier science is influenced by personal or social values that increase its subjectivity, while core science is the least subjective type of scientific knowledge (Longino, 1990). In other words, ES are value-laden by subjectivity to a degree, but the degree depends on whether frontier science or core science is being practised.

Values can be classified into three groups (Snow, 1987):

1. Those belonging to scientific knowledge ("epistemic values"—the cognitive dimension), for example, trustworthy data, predictive accuracy, consistency with established ideas, simplicity, and repeatability;
2. Those belonging to the conduct of the scientific community ("community values"—the sociological dimension), for example, honesty, skepticism, originality, and rationality; and
3. Values held consciously or unconsciously ("social values"—the personal dimension), for example, personal values or ideologies (such as stewardship, sustainability, or power and dominion over nature), professional identities, and institutional and personal allegiances.

Some values are more subjective than others.

To summarize, Eurocentric sciences are communities of practice within paradigms, both theoretical and empirical paradigms. Theories influence what scientists observe and how they interpret those observations. ES are what scientists do, but what they do is so diverse that it is unreasonable to suggest they share a common set of methods, such as *the* scientific method. However, they do share a wide range of scientific values that contribute to the rigour of scientific investigations. This rigour is also strengthened by trying to be objective, although that ideal can never be attained for a number of reasons. Realistically, *the rigour of ES decreases subjectivity* as much as possible. In short, ES are communal, theory-laden, and value-laden.

To bridge the cultures of ES and Indigenous peoples effectively, a science teacher needs this realistic understanding of ES. He or she also needs to be aware of fundamental assumptions scientists make about reality and about how to know the physical world. These assumptions serve as foundations for scientific observing and thinking.

Fundamental Presuppositions

Scholars have searched for a coherent set of attributes that characterize a scientific worldview or a commonality among the various paradigms within Eurocentric sciences. Their investigations have uncovered underlying and often subconscious ideas about what reality is, how scientists know the world and what their ideas are based on, as well as ideas about the values guiding their work. These fundamental ideas unite most, but not all, scientists. Several of these ideas or presuppositions are introduced here, along with a few important exceptions that reinforce the enormous diversity within ES.

These presuppositions are placed in separate though related categories to facilitate later comparisons of ES and Indigenous knowledge. The fundamental presuppositions enrich our understanding of ES and offer clear evidence that Eurocentric sciences are anchored in culture.

It is important to note that professional success does not rely on scientists being consciously aware of the content discussed here (Chalmers, 1999; Wong & Hodson, 2009, 2010). Perhaps this is why such content is usually absent from university science courses. However, the content *is* found in most university science education courses because of its relevance to an informed science teacher who daily encounters students embracing presuppositions of the natural world very different from those in ES.

These presuppositions tend to be ideological in nature, and a few of them continue to be debated among scientists and other scholars. Throughout this book, *ideology* refers to a set of beliefs about the world and society that legitimize certain practices (Fourez, 1989). Universalism and pluralism are examples of two ideologies.

Nature Is Knowable

A simple fundamental presupposition states that nature is knowable. Scientific knowledge is usually comprised of generalized descriptions called *laws* and mechanistic explanations *theories*. Mechanistic explanations can be models or a series of cause–effect events. But, a hypothesis is either a generalized description or a mechanistic explanation offered as a guide for preliminary investigations. If accepted within a paradigm, a hypothesis can either become a law or a theory. For example, Charles's gas *law*, which states temperature affects gas pressure, was established prior to the general acceptance of the kinetic-molecular *theory* of gases. Theories do not become laws.

A scientific model acts as a metaphor for how the physical world works. A scientific metaphor was originally anchored in the language or cultural ways of seeing the world by the people who imagined the model in the first place (Dyck, 1998). As a consequence, nature is knowable but that knowledge is necessarily imbued with metaphor (Goldstein & Goldstein, 1981). For example, the equation $F = ma$ is a culture-specific metaphor in the sense that it can only have meaning in a culture that practises mathematical quantification the way Eurocentric sciences do. Knowledge generated by scientists is largely dependent on metaphor.

In Euro-American cultures, mystery in nature tends to create a need to know nature, which in turn leads to investigations aimed at solving the mystery through scientific descriptions and explanations. The *elucidation of mystery* is an important intellectual goal for scientists but not for most Indigenous peoples, as shown in Chapter 6, the section "Mysterious."

Eurocentric Sciences Are Embedded in Social Contexts

The desire to know nature motivates scientific curiosity. The acquisition of knowledge to satisfy this scientific curiosity is a normal psychological goal for individual scientists, whether they work commercially or in "pure" academic research labs where unfettered curiosity results in knowledge for the purpose of acquiring knowledge. As described in Chapter 3, this scientific goal was established by natural philosophers when the Royal Society was institutionalized in the seventeenth century. At that time, the act of adhering to the value of knowledge for knowledge's sake allowed natural philosophers to avoid political disagreements with the church and royalty.

The value of knowledge for knowledge's sake has decreased in contemporary society. Most professional scientists are paid to generate, transform, or apply knowledge for the purpose of benefiting the commercial or government institutions they work for (Ziman, 1984). Examples of "pure" scientific research these days include the theoretical investigations into the origin of the universe conducted at Perimeter Institute for Theoretical Physics in Waterloo, Ontario, as well as government research institutes located worldwide.

R&D is now the conventional domain for most scientific research. For example, at NASA one finds "pure" research into the solar system and beyond, as well as R&D problem-solving for space exploration. The social goal of knowledge for knowledge's sake is now mostly absent in R&D, a fact lamented by some R&D scientists who would prefer to engage in more "pure" Eurocentric science.

Social science research conducted by sociologists, historians, anthropologists, and science educators has investigated the ways in which scientists work on a day-to-day basis. This research has uncovered many other social goals (Kelly, Carlsen, & Cunningham, 1993; Shapin, 2010; Wong & Hodson, 2010). Individually, scientists are motivated for several reasons: to satisfy their curiosity (either an *unfettered* one or one *targeted* by corporations or governments); to win research grants; to get their findings published or patents approved; to acquire or maintain credibility among their peers; to receive reasonable financial remuneration; and, in a few cases, to gain fame and fortune. Collectively, scientists employed by business, industry, government agencies, private foundations, and so on are necessarily guided by social goals that vary according to the nature of the projects in those institutions. Individual scientists may not personally aspire to a social goal held by their employer or by the group funding their research, but by their employment, they nevertheless are associated with social goals. These include material and economic progress, medical advances, and corporate profits (Shapin, 2010; Ziman, 2000). Because R&D always takes place in social contexts (Ziman, 1984), these social goals are unavoidable presuppositions in ES (Glasson, Frykholm, Mhango, & Phiri, 2006).

As with any employment, social contexts tend to express ideologies. In his book *Biology as Ideology: The Doctrine of DNA*, Harvard geneticist Richard Lewontin (1991) illustrates one role of ideology by describing Eurocentric science as a social institution. He outlines a dual process in which ideologies are indirectly expressed through Eurocentric science: some social institutions influence or control what scientists do and say; and some "use what scientists do and say to further support the institutions of society" (p. 4).

One feature common to most social goals of ES is competition. This has positive consequences because it accelerates progress and encourages a critical examination of new ideas. But, it also has at least three negative effects: first, when it can cause scientists to be secretive and not share their results; second, when it encourages them to quit their profession due to the stress of competition; and third, when it leads them, on rare occasions, to deliberately fabricate or misrepresent results to get ahead (Wong & Hodson, 2010). Of course, similar negative aspects can apply to competition in society as a whole.

Competitive behaviour within ES can alienate students whose self-identities revolve around co-operation. Cultural differences between students' self-identities and their perceptions of the social goals of ES may underlie the negative way some students react to ES in school science. Such science-shy students resist learning conventional school science in a meaningful way (Aikenhead, 2006; Taconis & Kessels, 2009). However, group work that fosters collaboration is often inviting for these students.

Eurocentric Sciences Have Predictive Validity

A primary aim of any scientific investigation is to test the predictability of an idea that describes or explains a phenomenon. This is a major strength of the ES approach. Hypotheses stand or fall on their ability to predict the outcomes of experiments accurately. When the predictability of a scientific law or theory is shown to be inaccurate by important contradictory evidence, a paradigm's acceptability is threatened.

In addition to experiments, there are other types of empirical research that include observational, field-based, and quality-control studies. All contribute to ES, depending on the paradigm. Their descriptions of the physical world fit into a scientific way of collecting data to help explain or predict how, for example, an ecosystem works. Such ecosystem descriptions are valued for their predictable broad applicability to similar ecosystems and for their success at initiating future experimental research. This, in turn, is all about identifying successful predictors. Predictive validity is a foundational presupposition for Eurocentric sciences.

However, logically speaking, false ideas can sometimes lead to accurate predictions. For example, Ptolemy's geocentric model of the solar system—or more precisely, his geostatic model—is very good at making important predictions, such as the timing of eclipses.

Predictive validity can be contrasted with *content* validity in which the content is true in some absolute sense. Scientific predictive validity looks at *how* the universe works. Content validity, on the other hand, addresses *what* the universe *is*, something Aristotle called *intelligible essences*. These essences are described by Battiste and Henderson (2000) as follows:

> The essence is the form of matter that lends each being its distinctive identity. The supporters of the doctrine of intelligible essences held that the standards of right and wrong must also have 'essences' that thought can comprehend. Plato's ethics and St. Thomas Aquinas's theory of natural law exemplify this line of argument. (p. 121)

According to Aristotle, if we deny the validity of intelligible essences—that which makes content true—we are denying *fidelity to a true world*. The issue here is what is truth? We do not offer an answer, but Aristotle's idea of intelligible essences resurfaces later in Chapter 6 (the section "Valid").

Consciously or unconsciously, scientists limit the validity of their knowledge to its ability to predict. This has resulted in the best representation of "truth" that a community of scientists can offer. But, this knowledge is restricted to describing and explaining *how* the universe works. Discussion on this subject continues in the section "Reality Is Reproduced or Represented by Scientific Knowledge" later in this chapter.

The ability to predict and its association with the ability to control phenomena are also described later in this chapter.

Scientific Knowledge Is Dynamic

Knowledge produced by scientists is open to change. It is dynamic and therefore tentative. John Polanyi (2009) takes this characteristic a step further when writing about scientific laws:

> This serves as a reminder that the Creator did not originate them; human beings did. Human constructs though they are, they give evidence of their power by opening a Pandora's box of possibilities.... Our fundamental beliefs—about matter, motion, life and the cosmos—have had to be revolutionized several times. Our [scientific] community has emerged intact and strengthened. (A15)

The power of these human constructs arises from their predictive validity, of course. The refinement of ideas on the basis of new ideas or evidence is a strength of ES. But, human constructs generated by scientists are also influenced by scientists' language, education, experiences, intuition, culture, and paradigm (Goldstein & Goldstein, 1981; Holton, 1978). Again, Shapin (2010) goes further, subtitling his book *Historical studies of science as if it was produced by people with bodies, situated in time, space, culture, and society, and struggling for credibility and authority*. All of these elements add to the tentative nature of scientific knowledge. Even fundamental ideas in ES are susceptible to change, when one paradigm replaces another.

The dynamic character of scientific knowledge is shared to some extent with Indigenous knowledge, as shown in Chapter 6, the section "Dynamic."

Scientific Knowledge Is Generalizable

A fundamental ES assumption is that scientific concepts apply consistently over time and anywhere in the universe. "Correct" scientific results can be reproduced or verified by any scientist worldwide. It is unacceptable to have inconsistent scientific thinking or experimental results. Sir Isaac Newton expressed this same idea in his "Third Rule of Reasoning" of natural philosophy, stated in his 1687 book, *Principia* (Cajori, 1962, p. 398): "The qualities of bodies … which are found to belong to all bodies within the reach of our experiments are to be esteemed the universal qualities of all bodies whatsoever." Scientists rely on the generalizability of empirical evidence and scientific concepts. Generalizability gives scientific knowledge power, especially the power to predict and control.

It is easy to confuse generalizability with universalism (see Chapter 3, the section "The Problem with Universalism"). Generalizability is part of a cluster of assumptions, which together are called "universalism." Universalism is a broader and more complex idea than generalizability.

Generalizability does hold a downside for science teachers. Generalizable knowledge expresses an idealized form that is decontextualized and abstracted from personal experience (Hodson, 2009, Ch. 8). For example, anyone who dares ride a bicycle strictly according to Newton's laws of motion will painfully be reminded that his laws ignore friction, one of the practicalities of everyday life. This idealized, academic, scientific knowledge can be contrasted with the more practical scientific knowledge used in the everyday world of science-related occupations and science-related public events or issues (Aikenhead, 2006). For example, "science-in-action" occurs in a hospital where nurses constantly deal with science-related medical events (Aikenhead, 2005). Except for the few nurses who have a scientific worldview and were science-oriented students in school, most nurses normally use their professional knowledge to make workplace decisions.

A nurse can use science-in-action content to conclude that a patient's symptoms may indicate stress due to lack of oxygen. This content may at one time have been *deconstructed* from idealized academic content concerning gases, liquids, partial pressures, and physiology, and then *reconstructed* to fit the specific situation of a surgical ward (Aikenhead, 2005). Some academic science content was transformed into science-in-action content for nurses. Even though the academic knowledge formed the basis of understanding for theoretical relationships concerning oxygen deprivation, gas pressures, stress, and so on, in the busy practical world of a hospital ward, it is the use of science-in-action content that is appropriate.

It is not surprising that a large majority of science-shy students resist learning conventional school science that is mainly abstract, out of context, and seemingly impractical. These students find it difficult and often impossible to deconstruct and then reconstruct abstract scientific concepts for use in concrete everyday settings. However, science-oriented students usually feel very comfortable with abstract and decontextualized knowledge. Their curiosity is often piqued by the problem-solving task of deconstruction, reconstruction, and integration into an everyday context. Most science teachers, of course, were at one time science-oriented students.

Even though Eurocentric science-in-action is largely based in a unique context such as an everyday event, its generalizability across similar contexts is highly valued. A very different presupposition is found in Indigenous knowledge (see Chapter 6, the section "Place-Based").

Eurocentric Sciences Operate on Rectilinear Time

The presupposition of rectilinear time is a Eurocentric concept and the most commonly held idea of time today. It consists of the notion that time moves in a uniformly and limitlessly linear way from past to present to the future, and that each unit of time is always the same arbitrary mathematical quantity.

Bolter (1984) analyzes three different concepts of time: those held by the ancient Greeks, by Europeans in the Renaissance period, and by modern computer engineers. He concludes that a particular technology specific to each culture defines three different concepts of time. For ancient Greece it was the production of clay pots, for fourteenth-century Europe the mechanical clock, and for modern times possibly the computer. Of interest here is Bolter's contention that the concept of rectilinear time was constructed as a result of the mechanical clock.

> What kind of a universe did the clock suggest? A precise and ordered cosmos, for the clockwork divided time into arbitrary, mathematical units. It encouraged men [sic] to abstract and quantify their experience of time, and it was this process of abstraction that led to the creation of modern astronomy and physics in later centuries.... The clock made explicit a view of the universe that orthodox Christianity had been tacitly encouraging for centuries. (p. 27)

Before the availability of mechanical clocks, the general public's view of time was a subjective personal concept. Sundials, water clocks, and hourglasses did little to change this subjective view of time because these devices simply indicated arbitrary, discrete amounts of time. Time intervals varied according to the season or the specific measuring device.

Bolter speculates that today we may be living with a radically new defining technology—the low-temperature CPU computer, in which the concept of time surpasses twenty-first-century human experience with time. Whereas "an ordinary clock produces only a series of identical seconds, minutes, and hours, a computer transforms seconds or microseconds or nanoseconds into information. The enormous speed of this transformation puts the computer's operation in a temporal world that is outside of human experience" (pp. 102–103).

Bolter neglects to consider another well-established idea of time—the cyclical concept held by most Indigenous peoples the world over. This concept of time is further examined in Chapter 6 (the section "Based on Cyclical Time").

For now, most of ES embrace rectilinear time as an absolute feature of reality. Paradigms that draw on the general and special theories of relativity are exceptions (Greene, 1999).

Eurocentric Sciences Subscribe to Cartesian Dualism

A fundamental idea about the universe took root in Renaissance Europe and became well-known when René Descartes wrote about it in 1641. The idea, known as 'Cartesian dualism' today, divides existence into two parts: matter and mind. The natural world of interest to most scientists is that of matter and energy only—the material world that occupies space. Seen this way, matter is necessarily devoid of any kind of human intuition, spiritual force, or divinity because these elements belong to the mind—the non-material world that does not take up space. In Cartesian dualism, the material and non-material worlds, sometimes referred to as physical and metaphysical worlds, are distinct, independent, and non-interacting.

People who do not subscribe to Cartesian dualism (and they include most of the world's population) often perceive dualism as something that destroys their unity of existence (Irzik, 1998). Their alternative to Cartesian dualism is *monism*, in which matter and mind intermingle and interact with each other (see Chapter 6, the section "Monist").

Within ES, few paradigms follow a monist view of the world. One that does is particle physics, described by Malaysian physicist Seng Piew Loo (2007, p. 105) as "the marriage between quantum physics and cosmology." This monist paradigm enjoys a history of public attention as seen, for example, in Zukav's (1979) *The Dancing Wu Li Masters*. According to Loo (2007), an "Eastern monist" perspective has replaced a Cartesian "Occidental dualist" perspective in the field of particle physics. In addition to that field, certain other paradigms have also adopted a monist presupposition—within quantum theory (Duran, 2007), ecology (Capra, 1996; Menzies, 2006; Pierotti & Wildcat, 1997; Turner et al., 2000; Worster, 1994), and sustainability science (Clark & Dickson, 2003).

Most scientific paradigms, however, are tied to Cartesian duality.

Eurocentric Sciences Are Reductionist

Reductionism describes a general approach to scientific thinking. It is opposite to holistic thinking favoured by Indigenous peoples. Reductionism assumes that scientists can understand "the structure and function of the whole in terms of the structure and function of its parts" (Irzik, 1998, p. 168). Most scientists analyze or mentally break apart a complex phenomenon into simple parts, factors, or variables that can be measured, conceptualized, and experimented with (Dyck, 1998). They investigate the parts to try to explain how the whole thing works. They assume the "whole" can be understood by putting together information about the parts (Ziman, 1984). A Canadian Indigenous Elder once defined scientists as people who take things apart to find the centre (Blackwater, 2009).

A small number of scientists hold a more holistic belief about knowing nature. These scientists work in areas such as ecology (Capra, 1996; Pierotti & Wildcat, 1997), sustainability science (Clark & Dickson, 2003), geology (Glasson et al., 2006), emergent complex systems or chaos theory (Hazen, 2005), and particle physics (Loo, 2007; Zukav, 1979). Many of these non-reductionist Eurocentric fields embrace Cartesian dualism, but a few do not and are comprised of monist thinkers.

Eurocentric Sciences Are Anthropocentric

Anthropocentrism establishes a hierarchy of importance in which people have a special status within nature—above that of animals, plants, and non-living things in nature. As a result, nature can be seen as a servant to humankind. Anthropocentrism is sanctioned by some religious and philosophical doctrines, in particular by the Judeo-Christian tradition (Cajete, 2000b; Reiss, 2008). The scheme was embraced by seventeenth-century natural philosophers and enthusiastically supported by nineteenth-century professional scientists, who were then free to investigate, rule, and exploit nature through the divine sanction of Christianity (Mendelsohn, 1976).

Anthropocentrism creates a dichotomy of 'humankind versus nature,' which in turn conveys the value or ideology 'power and dominion over nature' (Cajete, 2006; Mendelsohn, 1976) and the acceptability of using aspects of nature to further certain social objectives. In pharmacology, for example, animal studies precede human studies; while in electricity production, the natural power of flowing water is harnessed. As a result, the public may regard scientists as manipulators of nature because of their experiments on nature and their participation in projects that have altered nature. The explosion of the first atomic weapon tested in the desert at Alamogordo, New Mexico, on July 16, 1945, has since served as a dramatic icon of the control over nature associated with ES, despite such eminent physicists as Albert Einstein disagreeing with its application. However, for some scientific fields such as astronomy and deep ecology, as well as some other specific research projects, anthropocentrism is not an issue.

Students and most Indigenous peoples who believe that humans enjoy a non-hierarchical and egalitarian relationship with nature may find anthropocentrism distasteful. These students, Indigenous and non-Indigenous, tend to resist attempts to teach them school science (Cobern, 2000). This is another case of a cultural clash between a student and school science that results in indifference or negative reactions against school science. Science-shy students are seldom able to articulate reasons for their reaction other than saying, "It's a foreign culture" or "I don't like the morality of scientists." Chapter 6 looks at an egalitarian relationship with nature commonly embraced by Indigenous communities.

The Material World Is Governed by Quantification

The presupposition of quantification assumes that the material world is governed by objective mathematical relationships. Theoretical physicists are prone to say 'the language of nature is differential equations.' As a result, the quantification of natural phenomena is to be expected in ES. Although some ES are not known for their quantification, their status as a discipline within the larger scientific community varies accordingly. Concepts such as the complexity of life, for instance, are not considered scientific unless they are measurable (Hazen, 2005).

By representing things, events, and people by numbers, the presupposition of quantification encourages some scientists and others to believe that they can objectify a thing, event, or person by stripping them of their qualitative, human, or spiritual attributes—their intelligible essences. The quantification of things, events, and people occurs within the framework of Cartesian duality.

We gain further insight into the quantification presupposition by remembering that mathematical relationships are also conveyed through geometry. Geometry, as you may recall from high school, is built upon a small number of axioms or ideas that are neither true nor false, but are simply assumed. The Euclidean geometry system, developed in ancient Greece, is integral to Euro-American cultures. One Euclidean axiom of interest here concerns parallel lines: If two straight lines, A and B, are both at right angles to a third straight line, C, then lines A and B are parallel and never meet, except perhaps at infinity. In a Euclidean worldview, as Newton's was in the seventeenth century, the universe is perceived and conceived mathematically through the eyes of Euclid. In the nineteenth century, however, mathematicians such as Riemann invented geometry systems different from Euclid's by altering one or more of the Euclidean axioms. Parallel lines *do* meet in a Riemannian geometric world of curved space, in contrast with Euclidean flat space. Lines A and B, like the lines of longitude on a globe, are at right angles to line C, the equator, and so they do meet—at the north or south poles. Interestingly, Einstein had to use a non-Euclidean geometry system to develop his relativity theory's space-time continuum (Greene, 1999). To go from a Newtonian paradigm to an Einsteinian paradigm, a dramatic change in one's geometric thinking is required. Other cultures have also developed geometric systems of points, lines, and planes as in China about 400 BCE (Needham, 1956).

We draw four conclusions here: (1) the two-dimensional Greek world of Euclid is an idealized abstract way of thinking about the world, with many practical uses; (2) the practicality of the Euclidean system today has made it attractive to most nations worldwide, where it has been either freely imported by a nation or exported by a colonizing nation; (3) Euclidean geometry is taught in schools the world over, not because it is reality, but because it is useful, despite its Greek-based, Euro-American cultural bias; and most importantly, (4) quantification also reflects a *cultural bias* in ES, as do all the other presuppositions discussed in this chapter.

Because quantification depersonalizes the knowledge that scientists generate, it perpetuates the myth that scientists achieve objectivity through quantification. The objectification of things by quantifying them is an aspect of ES that some students may find inappropriate. For one thing, it turns their personal world into an impersonal world; for another, students feel vulnerable to its process because they do not want to be objectified themselves. It might be helpful to such students for the teacher to point out that some quantitative measures, such as the concentration of glucose in the blood, can be of paramount importance to the health and survival of a person if it leads to treatment for a medical condition such as diabetes. The power of quantified measures is not reduced by recognizing that measurements too are theory-laden observations (Goldstein & Goldstein, 1981).

In the culture of the ES, there is a tendency to believe that if something can be measured or quantified, it must exist as reality, rather than as a conceptual invention that metaphorically represents reality. For example, if psychologists gather IQ scores, this does not mean an IQ actually exists in a person. In this way, quantification reifies ideas or attempts to make them concrete by considering "the products of human activity as if they were something else than human products—such as facts of nature" (Berger & Luckmann, 1966, p. 89). By association, quantification also tends to reify scientists' observations, concepts, descriptions, and mechanistic explanations. Reification leads directly to the presupposition of realism.

Reality Is Reproduced or Represented by Scientific Knowledge

Do unobservable things, such as gravity, the helical structure of DNA, and neutrons, referred to in scientific laws, theories, and models, actually exist? Different answers to that question have evolved over time, expressed as realism, critical realism, instrumentalism, and constructivism. Each will be described in turn because these viewpoints have a direct bearing on how easily a science teacher can bridge scientific and Indigenous cultures.

A presupposition of *realism* suggests that unobservable things exist—they are real. According to realists, when scientific logic is applied to one's senses or to observations obtained from an instrument, the result is a true and objective image of reality (Ziman, 1984). In other words, realism postulates that scientists describe reality without the influence of their very act of perceiving. The resulting knowledge of nature is therefore thought to be a true picture of things as they really are, rather than a *representation* of reality tempered by human perception, imagination, social conventions, language, and cultural presuppositions. A representation of reality, according to realists, would fall short of their ideal goal of absolute objectivity. Some realists allow that their knowledge is only a 'best approximation'—almost a true image. This compromise is known as *critical realism* (Hodson, 2009).

A belief in realism or critical realism can be challenged on a number of grounds:

- Scientific knowledge is largely based on metaphor (see the section "Nature is Knowable"), thus it only *represents* the real world; it is not the real world or not necessarily a close approximation.
- Consider that 'scientific truth' is the end result of gathering valid evidence, arguing over that evidence and its interpretation, and finally, reaching a consensus among a group of scientists on what to believe. Each of these activities occurs in a human context, not in 'the Creator's domain' to paraphrase Polanyi from his quotation earlier in this chapter.
- Revolutionary changes from one paradigm to another, as seen in the movement from a Newtonian to an Einsteinian paradigm, can cause dramatic changes to scientific truth, and to "our fundamental beliefs—about matter, motion, life and the cosmos" (taken from that same Polanyi quote). Revolutionary paradigm changes seem to deny an ever increasing best approximation of reality held by critical realists.

Therefore, it is reasonable to conclude that scientific laws, theories, and models *represent* reality, a reality mediated by human perception, imagination, social conventions, language, and cultural presuppositions. Have realists mistaken abstract idealized concepts for reality?

Realists claim fidelity of a true world by reifying scientific observations. This may sound like they are talking about intelligible essences, but they are not. Realism pertains only to the Cartesian *material* world, while intelligible essences pertain to both the material and non-material worlds.

Realism tends to be conveyed by school science resources, the public media, and university science and engineering courses. Its appearance in school science has been called "naïve realism" (Milne & Taylor, 1998;

Nadeau & Désautels, 1984; Yore, Hand, & Florence, 2004). For example, many people, including some physics teachers among them, continue to believe that Newton's concept of gravity is real. They think that gravity actually exists. However, when the predictability of gravity seriously failed about 100 years ago by not accounting for Mercury's orbit around the Sun, a shift from a Newtonian paradigm to an Einsteinian paradigm ensued. This was accompanied by a shift from Euclidean-based thinking to non-Euclidian thinking in which the concept of warped space-time replaced the concept of gravity. Gravity is an outdated idea, although it continues to be very useful to many scientists and engineers, and it is what people who have been taught about gravity think they are experiencing. Because reality is never out of date, gravity could not have been part of reality. To think otherwise sounds naïve.

Realism or critical realism is a component of universalism (Chapter 3), which "logically" discounts the legitimacy of Indigenous knowledge by presuming a monopoly on the description of physical reality. As a consequence of this, realism or critical realism undermines bridges between a scientific worldview and Indigenous knowledge.

Realism or critical realism is certainly not the only presupposition embraced by scientists. Another viewpoint states that gravity is simply a conceptual tool in a physicist's intellectual tool kit. Theoretical entities such as gravity, the helical structure of DNA, and neutrons are tools or instruments "for making empirical predictions and achieving other practical ends" (Stanford, 2006, p. 400). Scientific ideas are seen as instruments of description, explanation, and prediction.

This presupposition is known as *instrumentalism* or "conventionalism" (Ziman, 1984). Scientific laws, theories, and models serve as intellectual instruments or cultural conventions to help humans predict what will likely occur (predictive validity), rather than describe what the world really is (content validity). Some scientists move between realism and instrumentalism depending on the task they perform in their work (Wong & Hodson, 2009).

Another alternative presupposition to realism or critical realism goes one step further by recognizing the important contributions made by social contexts in which paradigms and scientific employment exist. The concepts that scientists talk about originally came into being through a series of social interactions associated with acquiring empirical data, interpreting the data, crafting an argument to persuade other scientists to accept the research team's interpretation, and, finally, engaging in consensus making to help determine what will be "true" at the present time (Kelly et al., 1993). Evidence-based theoretical entities are socially constructed by scientists engaged in a series of processes ending with consensus making as in the peer-review process. This presupposition is known as *social constructivism*. A reliance on empirical data continues to be an important value in social constructivism. Social constructivism does not ignore reliable and valid evidence, as some realists would argue.

Both instrumentalism and social constructivism propose that scientific laws, theories, and models are *invented* by scientists based on empirical data but guided by assumptions, prior knowledge, creativity, and paradigm allegiances. In contrast, realism and critical realism suggest that scientific laws, theories, and models are *discovered* by scientists in their empirical data, in the same sense that one discovers gold. The results are therefore assumed to be a true, or almost true, picture of reality.

All four presuppositions—realism, critical realism, instrumentalism, and social constructivism—are rejected by a type of postmodern view about scientists and reality, a viewpoint called "cultural-historical activity theory" (van Eijck & Roth, 2007, p. 934), in which "knowledge is a dynamic set of artifacts that simultaneously mediate activity and are produced by activity.... [It] is always *knowledge in context*... [The theory] emphasizes cultural-historical factors and interactions between subjects and context." According to this theory, knowledge is not in one's head, but is achieved through a group activity.

Empirical Data Speak for Themselves: Positivism

With its roots in the early twentieth century (Holton, 1978) and sustained by an ideology of technical rationality (Habermas, 1972), the presupposition of positivism exerted a strong influence on how people think about ES, up until the 1960s (Ziman, 1984). Positivism passionately emphasizes logic applied impartially to theory-neutral observations and to strict empirical and experimental methodologies, all of which yield objective, value-free, universal, culture-free, secure knowledge of nature. This description is at odds with that of ES in this book. Positivism's emphasis on logical procedures has kept alive the myth of *the* scientific method. Positivists are invariably realists and will contend that empirical data speak for themselves. Once an experimental observation is "discovered," the research conclusion is logically straightforward, unique, and free from a scientist's subjectivity.

Consequently, the research conclusion is free from cultural influences such as worldviews, presuppositions, and allegiances to a paradigm or to a particular funding agency. Positivism is also known as logical positivism.

Ironically, the positivist agenda was to construct a science *free from* any worldview, presupposition, or ideology. This was in spite of the fact that positivism follows a particular ideology itself, a set of beliefs about the world and society that legitimizes certain social practices. For instance, positivists consider their scientific thinking to be the ultimate measure of rationality (Holton, 1978), and they therefore believe their knowledge represents the fidelity of a true world—an ideology in itself. Positivism embodies universalism in which there can be only one ideal, one norm, and one standard—that of the positivist. From this, it is simply common sense for a positivist to regard all other knowledge systems and ways of knowing nature as inferior, Indigenous knowledge included. Positivists believe "that everything that is real is scientifically provable, so what can not be scientifically proven is not real" (Lyver, Jones, & Moller, 2009). In essence, positivism expresses Eurocentric superiority over all other cultures and celebrates the value of power and dominion over nature.

Over the last several decades, scientists and scholars have generally moved away from a positivist ideology. For example, Albert Einstein wrote, "The sense-experiences are the given subject-matter. But the theory that shall interpret them is man-made" (1956, p. 98). The flight from positivism accelerated after the first edition of Kuhn's book, *The Structure of Scientific Revolutions*, in 1962. Rather than being replaced by another presupposition as was realism, positivism is simply disappearing. It is out of date.

Indigenous neuroscientist Lillian Dyck (1998, p. 88) mentions how scientists "create the illusion" that they are objectively removed from their scientific phenomena. The illusion is conveyed by:

- the communication styles they employ; for example, the third person and passive voice;
- a positivist point of view conveyed in their textbooks;
- the type of training they receive to become scientists (university science programs);
- the concealment of intuitive creativity that helps solve a scientific problem; for example, a nighttime dream or voices from nature may inspire a way "to design an experiment to provide scientifically acceptable evidence" (p. 96).

Strong vestiges of positivism still remain in society. We hear people say, for instance, "Only the hard sciences can tell us what the facts are."

Conclusion

There is much diversity within Eurocentric sciences—in vastly different paradigms, in the variety of workplaces, and in the many types of scientists. This diversity defies any characterization of ES as having a single methodology, a single purpose, and a single worldview. Every scientist is unique in some way. Accordingly, this chapter has examined several myths and stereotypes—the typical scientist; *the* scientific method; the attainment of objectivity; realism and critical realism; and the existence of value-free or culture-free observations, hypotheses, laws, theories, and models.

On the other hand, descriptions of ES have revealed one common underlying feature of Eurocentric sciences: *They are all culture-based human communities*. Scientific disciplines and their paradigms are subcultures within the culture of Eurocentric science (Pickering, 1992; Sillitoe, 2007; Wong & Hodson, 2010). This tells us that ES are very much a human endeavour in many different ways (McClune & Jarman, 2010; Goldstein & Goldstein, 1981; Shapin, 2010; Ziman, 1984). Because the scientific enterprise originated and evolved mostly within Euro-American cultures, scientists mainly embrace a Eurocentric worldview in their ways of knowing, metaphysics, and the value systems used when they engage in scientific professional work. Specifically, scientists embrace variations on a scientific worldview described throughout this chapter.

Educational research unequivocally shows that few students and adults are aware of the many human dimensions to the ES—how scientists behave within paradigms, their multiple methodologies, their culture-laden presuppositions, and their social and economic contexts of employment (Aikenhead, 2006). Whether or not students and adults should be familiar with such knowledge in order to be scientifically literate is an issue beyond the scope of this book.

We simply wish to make two points. First, conventional school science, the media, and university undergraduate science programs have not conveyed realistic ideas about the scientific enterprise. By concentrating on its content, methods, and techniques, little or no attention was given to its public myths and stereotypes, its human dimensions, its social aspects, and its cultural features. Indeed, some may hold that the media and conventional science curricula in schools and universities have projected the antithesis of authenticity by promoting a stereotypic, idealized, solely academic science point of view (Hodson, 1998; Gaskell, 1992; Wong & Hodson, 2009, Yore et al., 2004). Perhaps this chapter has challenged some readers to reflect on their own ideas about ES and how those ideas were conveyed to them in science classrooms. Teaching what we were taught is a cycle that can be broken.

Our second point is that a human-oriented perspective on Eurocentric sciences as communities of practice, rather than a universalist content-oriented one, helps free our minds to be critically open to alternative ways of knowing nature.

A human-oriented perspective does not mean that scientific results are *merely* social constructions, purely arbitrary or highly subjective. On the contrary, all scientific conclusions must firmly rest on credible observations, no matter how theory-laden these may be. The essence of the ES was captured beautifully for the general public by Mary Budd Rowe (1995), an eminent American science educator, when she wrote, "Science is not just facts, but the meaning that people give to them—by weaving information into a story about how nature operates" (p. 181).

A feature common to ES is a particular set of foundational beliefs—presuppositions—upon which scientific observing and thinking are based. As strange as it may sound, scientists' success in their profession is unrelated to how accurately they can articulate these fundamental presuppositions (Chalmers, 1999; Wong & Hodson, 2009, 2010). Most scientists rarely have to think about them. That is why sociology, history, and anthropology researchers who study scientists have a less subjective approach to understanding fundamental presuppositions of the ES.

These cultural features of ES are summarized here. None of them devalues the power and worthiness of ES in today's world. They are simply the axioms of most Eurocentric sciences.

Eurocentric sciences are communal, but competitively so. Scientists embrace many values (cognitive, sociological, and personal values), a multitude of methods, and varying degrees of imaginative intuition. Many are awed by the beauty or elegance of their ideas and of nature. Scientists assume nature is knowable by elucidating its mysteries. Their knowledge system is generally characterized by the following:

- metaphor
- anthropocentrism
- predictive validity
- consensus making—a validation process that includes, for example, empirical evidence, rigorous methods, rational argumentation, peer review, and paradigm allegiances held by a group of scientists
- minimized subjectivity within a community of scientists
- tentativeness (highly dynamic for frontier science, much less so for core science)
- generalizability of idealized concepts
- rectilinear time
- Cartesian dualism (in almost all cases)
- reductionism (in most cases)
- quantification (in many cases).

Debate continues over

- the degree to which social goals influence what scientific "truth" will be;
- the credibility of realism, critical realism, instrumentalism, social constructivism, and cultural-historical activity theory;
- the appropriateness of positivism.

During the past 50 years since Thomas Kuhn's (1962) *The Structure of Scientific Revolutions*, many scholars, including some eminent scientists like John Ziman and John Polanyi, have recognized the clear influence of social goals on frontier science. They have disputed the credibility of realism or critical realism, replacing it with instrumentalism or social constructivism. And they have found positivism to be highly inappropriate. These developments are crucial to teachers who are preparing to include Indigenous knowledge in school science.

It is important to repeat this core idea: The presuppositions of realism, critical realism, and positivism *deny* the legitimacy of Indigenous knowledge as a foundational way of understanding the physical world. This denial trivializes Indigenous knowledge to the status of tokenism. The denial undermines attempts to *complement* a scientific worldview with an appreciation of an Indigenous worldview in school science. The denial also discourages many Indigenous students in science classes, thereby contributing to their under-representation in science-related occupations and, as a result, their diminished participation in their country's society.

This chapter offers a mindset for building *accurate, reliable,* and *appropriate* bridges to Indigenous ways of knowing nature described in the next two chapters. Our ultimate aim is for all students to learn the best of Indigenous and scientific ways of knowing nature.

CHAPTER 5

Indigenous Knowledge: Background

An exploration of Indigenous knowledge begins by clarifying four important terms: *Indigenous, knowledge, nature,* and *coming to know.* This clarification creates a step toward bridging Eurocentric sciences and the diverse knowledge systems of Indigenous peoples worldwide.

General expressions that might stereotype Indigenous peoples are avoided as much as possible in this book. It is a challenge, however, to write about Indigenous cultures using a Euro-American language because a language reflects a collective worldview. The four important terms examined in this chapter have different meanings in Indigenous and Euro-American collective worldviews.

Our role as authors is that of a facilitator, helping science teachers gain some understanding of Indigenous cultures to assist them in moving back and forth between the familiar culture of Eurocentric sciences (ES) and less familiar cultures of Indigenous peoples. Moving back and forth provides opportunities to bridge the two cultures. In this endeavour, everyone is a lifelong learner.

Clarification of *Indigenous*

Indigenous peoples[15], according to the United Nations, are the descendants of the first people to inhabit a locality and self-identify as members

---

[15] This book follows the convention that the term "Indigenous" applies worldwide and includes the First Nations, Inuit, and Métis peoples of Canada, who under Section 35 of the Canadian Constitution are encompassed by the term "Aboriginal."

of a collective. They are recognized by other groups or by state authorities, and they wish to affirm and perpetuate their cultural distinctiveness in spite of colonial subjugation and pressures to assimilate (Battiste & Henderson, 2000, pp. 61–64). Indigenous people generally share a collective politic of resistance arising from commonly shared experiences of oppression. Niezen (2003, p. 246) described their shared experiences as "marginalization, economic servitude, and sociocultural genocide." They also share a common vision of self-determination and cultural survival.

In keeping with this United Nations perspective, Elizabeth McKinley (2007), a Māori scholar and science educator, acknowledged different types of Indigenous peoples. These include those whose colonial settlers/invaders have become numerically dominant, such as the First Nations of Canada and the American Indians of the United States, and those in developing nations where colonial settlers/invaders never reached a majority but left a legacy of colonization as in African and some Asian nations. In addition, McKinley warned, "Indigeneity is a heterogeneous, complex concept that is contextually bound" (2007, p. 202). "Indigeneity" means the state of being Indigenous. The qualification "contextually bound" means there is no universal definition for "Indigenous." Indigenous peoples worldwide tend to reject a universal definition, for fear it might create an outsider-imposed Indigenous identity, thereby colonizing them all over again (Niezen, 2003).

Other types of indigeneity exist. In the book, *What Is Indigenous Knowledge?* (Semali & Kincheloe, 1999), the authors discuss the indigeneity of local knowledge of a particular ecosystem, such as knowledge held by long-resident farmers who may be of Euro-American ethnicity. Joe Kincheloe had acquired such knowledge as an impoverished white Appalachian boy in rural Tennessee. His local knowledge of nature appears to be Indigenous in its characteristics and specific details. However, Semali and Kincheloe concluded that such knowledge could not be considered Indigenous because Kincheloe's Appalachian knowledge of nature was held by a member of a privileged group—white Americans. The same reasoning applies to, for example, Canadian farmers of Euro-American descent who know the local land intimately. The important political criterion concerning privilege or oppression of the knowledge holder reflects the United Nation's perspective on the meaning of Indigenous.

But what about non-Eurocentric knowledge systems about nature held by socially and economically privileged nations such as China, Japan, and a number of Islamic nations? Their non-Eurocentric knowledge systems are characterized by a long-standing, intimate, and ecological knowledge of nature, yet their peoples have not been colonized, by and large. The United Nations' definition of Indigenous does not apply in those cases. Instead, these ways of knowing nature can be called "neo-indigenous," described in Chapter 3 (the section "Clarification of *Science*"). The neo-indigenous category includes non-Indigenous people such as Joe Kincheloe and any other non-Indigenous long-resident farmers who know their local land intimately.

Clarification of *Knowledge* and *Nature*

The noun "knowledge" does not translate easily into most Indigenous languages, in part because English is a noun-rich linguistic system while Indigenous languages are verb-rich. When translated into English, the corresponding Indigenous expression for "knowledge" often results in something like "ways of living" or "ways of being." Thus, it is appropriate to adopt the more authentic phrase "ways of living in nature" in place of "knowledge of nature." The phrase "scientific knowledge" fits the context of Eurocentric thinking, whereas the expression "ways of living in nature" generally fits an Indigenous context, although different communities may prefer different wording. Throughout the remainder of this book, we will therefore use the expression "Indigenous ways of living in nature" (IWLN), rather than "Indigenous knowledge."[16] When the term "Indigenous knowledge" appears here in this book, we intend it to mean IWLN.

Greg Cajete (2006), renowned Tewa Knowledge Keeper[17] of the Santa Clara Pueblo Nation, New Mexico, reminds us that the word "nature" has quite different meanings for Indigenous peoples than in its Euro-American context:

> Nature is not simply a collection of objects [and energy], but rather a dynamic, ever-flowing river of creation inseparable from our own perceptions. Nature

---

[16] Alternatively, some Indigenous scholars write "Indigenous knowledges" when communicating in English to convey a sense different from the normal Anglo sense of "knowledge."

[17] A Knowledge Keeper (Knowledge Holder or Knowledge Advisor or Knowledge Trustee) is a respected person to whom people go to gain help or understanding related to a specific issue or event. Such people are expected to pass this understanding on to the next generation. Usually, a Knowledge Keeper is a type of "specialist" who does not have the same revered status of an Elder. However, some Elders prefer to be called Knowledge Keepers.

is the creative center from which we and everything else have come and to which we always return. (p. 250)

The expression "Indigenous ways of living in nature" (IWLN) assumes Indigenous meanings for the word "nature." These meanings reflect a great diversity of contexts, such as Indigenous people in both rural and in urban settings. The word "nature" in the context of ES refers to the universe's collection of matter and energy—the natural world.

Within every urban locale exist natural settings like creeks, rivers, and green spaces. And underneath every urban locale is nature or the land that gave rise to an Indigenous worldview. Urban students need to be familiar with the ways in which an urban environment is still connected to the natural world (Environics Institute, 2010). It is a matter of perspective. For example, concrete comes from rocks (called "the grandfathers" by some Indigenous people) and wood comes from forests. Furthermore, urban land was originally inhabited by a particular Indigenous nation at the time of European contact. This is another connection between urban students and the land.

Some Indigenous communities and writers prefer the phrase "Mother Earth" to the word "nature," but others do not; whether they are located in a rural or an urban setting. Those who prefer "nature" consider the phrase "Mother Earth" to be unduly influenced by European colonizers because it appears in Indigenous speech only after contact with the settlers/invaders, and particularly since the 1960s (CBC, 2003). Linguistically, "Mother Earth" can be traced back to Gaia, the Greek Earth Mother goddess, after whom the discipline of geology was named. The expression "Mother Earth" is not found in major First Nations and Inuit languages in Canada. However, some Indigenous people when expressing themselves in English do refer to "Mother Earth." The phrase evokes a much broader concept than "planet Earth" or "nature" by including spiritual aspects that sustain life, relationships, and responsibilities. Perhaps the phrase "Mother Earth" has been appropriated from the settlers/colonizers because it best conveys in English what Indigenous people understand to be a reverence for their relationship with all of creation on Earth. This creates common ground between them and the settlers/colonizers. Many Indigenous people use either the term "nature" or "Mother Earth," depending on the context. We shall do the same and continue to use the word "nature" in the expression "Indigenous ways of living in nature" (IWLN). In a science classroom, we suggest that a teacher use whichever term is preferred by the local Indigenous community, if the community has a preference. In urban multicultural classrooms, Indigenous students can still learn reverence for Mother Earth without living in nature beyond their urban setting.

Another exception to equating "nature" with "Mother Earth" comes from the Navajo (Diné) Nation in the United States where "nature" usually includes both Mother Earth and Father Sky (Maryboy & Begay, 1998). Some other Indigenous nations identify the physical universe that lies beyond Mother Earth in terms of Father Sky, Grandfather Sun, and Grandmother Moon.

The English expression "Indigenous knowledge" covertly conveys a Eurocentric noun-oriented way of thinking. It can make Indigenous people think in a Eurocentric way by having them accept a Eurocentric concept—knowledge—as suitable for their worldview. One method of colonizing Indigenous peoples is to replace their native language with English. The Eurocentric expression "Indigenous knowledge" subliminally forces Indigenous people to accept a Eurocentric perspective on understanding nature.

According to Cajete (1994, p. 87, italics in the original), "the word *Indigenous* ... means being so completely identified with a place that you reflect its very *entrails*, its soul."

The expression "Indigenous ways of living in nature" is plural—indicating numerous ways of living—because there are many Indigenous nations, tribes, and clans. Each group has its own unique way of living in nature and a slightly different collective worldview. In Canada, for example, there are diverse First Nations cultures each with their own traditions and ways of expressing themselves in English or French[18]. Each nation has its own specific ideas about reality — their way of knowing nature, how to understand the world, and what values those ideas are based on. For instance, Bastein (2004) wrote a book called *Blackfoot Ways of Knowing*. Similar books could be written for Cree, Dëne, Métis, Lakota, Mohawk, Oneida, Ojibwa, Mi'kmaw, and Inuit Nations. The phrase "Indigenous ways of living in nature" is a shortcut way of writing the unwieldy though more accurate phrase "a Cree or Dëne or Métis or Lakota or Mohawk or Oneida or Ojibwa or Mi'kmaw or Inuit, etc. way of living in nature."

When Indigenous ways of living in nature are described in this book, these descriptions do *not prescribe* what those ways should be. Instead, the

---

[18] English and French are the official languages in Canada.

descriptions *indicate* what might be expected of some Indigenous people in a community. Some ideas are shared across many Indigenous nations, tribes, and clans. But to know for sure, one must ask. By honestly asking, respect is shown for the local Indigenous way of living in nature. Each Indigenous group is different, but groups do share some general understandings.

For most Euro-American people, "knowledge" is a noun. It is something that can be given, taken, accumulated, banked, and assessed by paper and pencil examinations. Knowledge is something that exists independently of people; it exists separately from the knower. This common Eurocentric notion is totally foreign to most Indigenous worldviews. An Indigenous knower is intimately and personally interconnected with what it is they know. This is another reason that no equivalent word for "knowledge" exists in most Indigenous languages. The expression "Indigenous ways of living in nature" assumes this intimate connection between knower and what is known—the intermingling of mind and matter.

There is also an important connection between what is known and who shared that understanding. Therefore, one should always identify the Knowledge Keeper or Elder with the ideas learned from that person, and mention the permission received for sharing that understanding with others. This protocol conveys an authentic appreciation of IWLN and helps non-Indigenous people build cross-cultural bridges. Indigenous ways of living in nature are about *contemporary* times. Informed by generations of community-based wisdom, Indigenous people have adapted modern ideas and technology to sustain their goal of cultural survival. For instance, Indigenous communities are using Global Positioning System (GPS) devices to help record, analyze, and communicate information regarding their local resources without compromising their community's values, goals, and wisdom. Any reference to pre-contact IWLN should be specifically identified as such, for example, "historically traditional knowledge" or "pre-contact knowledge."

To conclude, scientific knowledge is noun-rich and impersonal, rather than action-rich and personal. For example, *Tsikala* (never sleeps) is the Kwa'k'wala word for "police." The tendency in ES is to view nature as a detached collection of matter and energy, rather than a web of relationships among everything in existence. This tendency of ES is one reason why many Indigenous students find scientific knowledge irrelevant to their action-oriented personal world of relationships (Cajete, 1999). To connect with Indigenous students, learning must become a personal part of who they are and what they do, in other words, their self-identity. These students' preferred standard of learning is not met by memorizing curriculum content. From an Indigenous perspective, learning is ideally about becoming whole, complete, and balanced in mental, spiritual, emotional, and physical ways. A balanced person can fulfill her or his responsibilities within the interdependent context of family, community, ceremonies, and nature relationships.

Clarification of *Coming to Know*

Meaningful learning for the Nehiyawak (Plains Cree people) is captured by the English phrase "coming to know" (Ermine, 1998), which means a Nehiyaw (a Plains Cree person) is on a quest to become wiser by living properly in his or her community and in nature. To live properly includes the *action* of living in harmony with the natural environment for the sake of the community's survival. This sense of learning is shared by most Indigenous communities. "Wisdom is based on generations of knowledge, close observation of natural order, and a cultural and spiritual consciousness articulated through traditional holistic language.... Another manifestation of Indigenous wisdom is active intelligence" (Maryboy, Begay, & Nichol, 2006, p. 9). In short, scientists pursue *knowledge* in an analytical and critical way, whereas Elders pursue *wisdom-in-action* as lifelong learning and as advice for a community's survival.

Knowledge and wisdom-in-action embody very different ways of understanding nature. Notable American Senator J. William Fulbright (1964) once wrote, "Science has radically changed the conditions of human life on Earth. It has expanded our knowledge and our power but not our capacity to use them with wisdom" (p. 5). Indigenous wisdom is intimately related to human action based on natural laws: "Nature provides a blue print of how to live well and all that is necessary to sustain life" (Michell, 2005, p. 39).

The process of *generating* Indigenous ways of living in nature (IWLN) is also "coming to know" (Cajete, 2000b), or "coming to knowing" (Peat, 1994), phrases that connote an ongoing journey. The process of *coming to know* differs from the Eurocentric science process *to know* with its emphasis on inventing ideas based on empirical evidence. Generating knowledge in ES implies a destination, such as a published record of a conclusion or an application for a patent. But coming to know is a journey toward wisdom-in-action, not a destination of discovery. In short, Elders *inhabit* the world while scientists *observe and interpret* the world.

This book cannot describe concrete details of Indigenous ways of living in nature. Unlike Eurocentric knowledge, coming to know a specific detail is an experiential journey that involves protocols, practice, and strengthening relationships. Ways of living in nature are action-oriented. "Woodlands Cree [Nîhîthewâk] epistemology is participatory, experiential, process-oriented, and ultimately spiritual" (Michell, 2005, p. 36). Reading this book will not provide an adequate experience in coming to know Indigenous ways of living in nature (IWLN), but it is an excellent place to begin *appreciating* what IWLN means. As mentioned earlier, IWLN cannot be given, taken, accumulated, banked, and assessed by paper and pencil examinations. IWLN must be experienced as follows:

- in a particular place in nature (it is place-based),
- in the context of multiple relationships with nature and people, and
- in the pursuit of wisdom-in-action for the purposes of survival.

In an urban setting, connections to the natural world can be made by learning where construction materials, such as wood and concrete, originate, and by learning the source of the animals and plants people eat. Because the disconnect between humans and nature is often greater in cities than in rural areas, teachers need to creatively bring nature into the classroom. Urban settings such as Toronto have abundant native plants in vacant lots as well as prolific wildlife such as raccoons, skunks, squirrels, and coyotes. These can all be potential subject matter to explore. As well, some parking lots might be 'reclaimed' by a community as a natural green space. City school systems need to bring students out to a more natural setting and into the land where the quality and quantity of learning escalates. As discussed in Chapter 8, students can learn *about* Indigenous ways of living in nature to some degree in a non-experiential way, in keeping with a science curriculum's content.

Indigenous ways of living in nature (IWLN) have been described in terms of competencies (Barnhardt & Kawagley, 2005):

> In Western terms, competency [in school science] is often assessed based on predetermined ideas of what a person should know, which is then measured indirectly through various forms of 'objective' tests. Such an approach does not address whether that person is actually capable of putting that knowledge into practice. In the traditional Native sense, competency has an unequivocal relationship to survival or extinction—if one fails as a caribou hunter, the

entire family is in jeopardy. One either has or does not have requisite knowledge [ways of living in nature], and it is tested in a real-world context. (p. 11)

In addition to experiential learning, IWLN are normally communicated and learned in the oral tradition. Indigenous languages are the main means of passing on IWLN. This is particularly helpful to Indigenous students who have not lost their mother tongue. It can also help them appreciate the richness of their own language even if they are unable to speak it fluently. Other methods of reinforcing or encouraging Indigenous worldviews within an Indigenous community include modelling the practices of others, listening to stories, singing songs, reciting prayers, dancing at celebrations, and participating in spiritual ceremonies. However, these methods are not usually appropriate for school classrooms. Oscar Kawagley and his colleagues (1998), scholars of the Yupiaq Nation in Alaska, point out that a Yupiaq way of knowing nature "is manifested most clearly in their technology" (p. 136). Technology includes processes and all sorts of artifacts, such as snowshoes, fashioned or manufactured for human purposes.

In contrast, school science is more typically communicated and learned in the written tradition in a way that conventionally pays little attention to such technology. IWLN can also be found in local entertainment such as drama, proverbs, and jokes (George, 1999). Proverbs are particularly powerful in the Indigenous cultures of Hawai'i (Chinn, 2008) and Aotearoa New Zealand (*Māori Proverbs*, 1992). The Māori *E kore e hekeheke he kakano rangatira* literally means, "I will never be lost, for I am of the seed of chiefs" (p. 22), but a clearer translation is, "Our ancestors will never die, for they live on in each of us" (p. 22).

In summary, coming to know is a personal, participatory, constructive process toward gaining wisdom-in-action. A book such as this is unable to provide concrete instances of Indigenous ways of living in nature (IWLN), such as how and when to pick 'rat root,' a plant in northern Saskatchewan that relieves sore throats and coughing. Instead, our book aims to prepare teachers' minds for learning IWLN as they are identified in a science curriculum and found in classroom resources. There is nothing like bringing the forest and land to Indigenous students, whether rural or urban, to give them pride in their identity and an oasis for spiritual rejuvenation.

With the benefit of a culturally sensitive understanding of key ideas discussed in this chapter, the next chapter considers specific fundamental attributes of IWLN—the Indigenous side of the cultural bridges being built by science teachers.

CHAPTER 6

Indigenous Ways of Living in Nature

"Traditional native knowledge" about the natural world is often extremely sophisticated and of considerable practical value (Suzuki, 2006, p. 12). Indigenous scholars have described their knowledge through ideas about what reality is, about how Indigenous peoples know the world and what these ideas are based on, and about the values that guide Indigenous communities worldwide. This chapter describes these concepts, giving voice to Indigenous scholars by quoting extensively from their work. More 'first-voice' sources, oral and written, can be found on the Internet, such as the Omushkego Oral History project (Bird, 2010) or the Four Directions Teaching project (National Indigenous Literacy Association, 2007).

We try to avoid describing a stereotypical pan-Indigenous way of knowing nature. Each clan, tribe, or nation has a unique set of understandings. Indigenous ways of living in nature (IWLN) are first and foremost local or *place-based,* they are not generalizable. Therefore, protocol dictates that the knowledge offered through the quotations cited in this chapter must be identified with the place or Indigenous nation associated with the writer. In this way, the reader is reminded that the validity of that knowledge may be restricted to that place.

Indigenous people often say, however, that while no two groups are the same, they all share commonalities in their own tribal ways of living in nature and in their Indigenous ways of understanding how to live. This chapter gives voice to these ideas held in common. Science teachers could draw upon them to initiate a conversation with their local reputable Knowledge Keepers to find out more about the local IWLN.

Fundamental Attributes

The following fundamental attributes of Indigenous ways of living in nature (IWLN) are essential to science teachers who are beginning to bridge Indigenous and scientific ways of knowing nature. IWLN are place-based, monist, holistic, relational, mysterious, dynamic, systematically empirical, based on cyclical time, valid, rational, and spiritual. These attributes comprise the essences of a collective Indigenous worldview. The topics are organized in a way that facilitates comparisons of IWLN and Eurocentric sciences (Chapter 7).

Place-Based

Unlike the generalizable type of knowledge of Eurocentric sciences (ES), Indigenous ways of living in nature (IWLN) are place-based (Michell et al., 2008). For example, while Eurocentric botanists use the Linnaean taxonomy system to identify plants anywhere on the planet, Elders employ their community's place-based traditions that "may include stories relating to [a plant's] use as a food source, its ceremonial uses, its complex preparation process, the traditional accounts of its use (as in purification ritual), its kin affiliations, and so on" (Snively & Corsiglia, 2001, p. 11). Sources of wood can be named according to the amount of heat given off when burned, or according to the technology for which they are used. Plants can also be named according to shape. For example, the strawberry is named *otehimin*—"a heart berry"—in Plains Cree because of its heart shape.

The quality of being place-based is both a strength and a limitation. Greg Cajete of the Santa Clara Pueblo Nation (2000b) describes some profound implications of a place-based reality:

> All human development is predicated on our interaction with the soil, the air, the climate, the plants, and the animals of the places in which we live. The inner archetypes in a place formed the spiritually based ecological mind-set required to establish and maintain a correct and sustainable relationship with place.... But people make a place as much as a place makes them. Native people interacted with the places in which they lived for such a long time that their landscapes became reflections of their very souls. (p. 187)

For example, the history of the Nîhîthewâk (Woodlands Cree) and the history of their land do not simply coexist; they are one and the same:

> We personify the Land—as our Mother Earth. It has memory.... To displace and disconnect Woodlands Cree people from the land is to sever the umbilical

cord and life-blood that nurtures an ancient way of life. Our Cree way of life requires that we maintain a balanced and interconnected relationship with the natural world. (Michell, 2005, p. 38)

Most Indigenous cultures worldwide share this attribute (Davis, 2009). Cajete (1999, p. 47) points out, "Native science[19] evolved in relationship to places and is therefore instilled with a 'sense of place.'"

Because Indigenous peoples' self-identities are imbued with such a strong sense of place, physical place becomes part of their spiritual space (Ermine, 1995). Their 'land as identity' differs dramatically from the commerce-oriented view of 'land as a commodity to be bought, depleted, and sold.' Cajete (2000b) wrote:

> Native peoples' places are sacred and bounded, and their science is used to understand, explain, and honor the life they are tied to in the greater circle of physical life. Sacred sites are mapped in the space of tribal memory to acknowledge forces that keep things in order and moving. The people learn to respect the life in the places they live, and thereby to preserve and perpetuate the ecology. (p. 77)

The concept of place is not necessarily restricted to the Eurocentric notion of land. Nancy Maryboy and her colleagues (2006) affirm:

> Like the Navajo, most Indigenous people are spiritually grounded in specific geophysical and celestial environments. Navajo cosmology is centered within their Four Sacred Mountains. Star constellations and other dynamically moving celestial objects visible from this location provide the natural holographic order underlying Navajo cosmography. (p. 7, Internet version)

In Saskatchewan, "When [Métis research] participants began talking about 'land,' they also included discussion about water, sun, moon, clouds, rain, and fire" (Michell et al., 2008, p. 96). Many of the creation stories in Indigenous cultures have teachings associated with the basic components of life, which include water, air, fire, and earth, or variations thereof, such as water, *wind*, fire, and *rocks*.

The place-based knowledge embedded in someone's cultural identity goes with that person wherever they may live, including in a big city. An Indigenous self-identity can be a resource to draw upon perhaps (see

---

[19] Cajete talks in terms of Native science, which we consider synonymous with Indigenous knowledge and IWLN. His expression "Native science" implies a pluralist view of science (see Chapter 3, the section "Clarification of *Science*").

Chapter 8, the section "Teacher Expectations") because it is not dependent on a person living in the natural setting of their ancestral territory.

Métis scholar Madeleine MacIvor (1995) offers an important suggestion for science teachers:

> There is no reason why the classroom must be contained within the school structure. By expanding the walls of the school into the community, the "Indian sense of place" is seemingly greatly expanded, as is the involvement of the community in the students' science education. (p. 89)

MacIvor's advice was supported by many Saskatchewan-based Indigenous people who participated in a research project about Indigenous knowledge in the science curriculum (Michell et al., 2008).

Monist

Earlier we described how Malaysian physicist Seng Piew Loo (2007) viewed monism as an alternative to Cartesian dualism (see Chapter 4, the section "Eurocentric Sciences Subscribe to Cartesian Dualism"). Monism mingles the material and non-material worlds, the physical and metaphysical. While most scientists view nature or the material world as comprised of inert matter and energy, Elders and most Indigenous people do not. Their reality could be described as a monist reality. However, they understand reality in a much deeper sense—nature is simultaneously material and *sacred*. Such a worldview is reinforced by the linguistic characteristics of Indigenous languages.

> Indigenous languages are, for the most part, verb-rich languages that are process- or action-oriented. They are generally aimed at describing "happenings" rather than objects. The languages of Indigenous peoples allow for the transcendence of boundaries. For example, the categorizing process in many Indigenous languages does not make use of dichotomies.... There is no animate/inanimate [living/nonliving] dichotomy. Everything is more or less animate [alive]. (Blackfoot Elder and scholar Little Bear, 2000, p. 78)

The material world is imbued with Spirit. Thus, everything in the universe is alive: animals, plants, humans, rocks, celestial bodies, natural forces, etc. (Battiste & Henderson, 2000; Cajete, 2000b; Kawagley et al., 1998). Native Alaskan Yupiaq scholars wrote:

> In Western science, the closest to Yupiaq science can be seen in the study of ecology, which incorporates biological, chemical, and physical systems (earth, air, fire, and water). However, even many ecologists have ignored the fifth

element, spirit. Lack of attention to the fifth element has resulted in a science that ignores the interaction and needs of societies and cultures within ecosystems. (Kawagley et al., 1998, p. 139)

The universe, including Mother Earth, encompasses both the physical and the spiritual simultaneously. (Indigenous spirituality is discussed in greater detail later in this chapter.)

Spirituality, which is not to be confused with a Euro-American concept of religion, was not part of the language of natural philosophy in sixteenth and seventeenth century Europe. Natural philosophers of that period wanted to avoid being seen as challenging religious and royal authorities. Natural philosophy was restricted to the material world. As a result, wisdom and environmental ethics became products of human reasoning *outside* scientific reasoning. For Elders, wisdom and environmental ethics are *within* the structure of nature itself (MacIvor, 1995). "Scientific knowledge and predictions of the consequences of human choices will inform people's decisions around how to act in wise and ethical ways, but science does not prescribe the wisdom itself" (Lyver et al., 2009). The biologist's concept of photosynthesis, for instance, does not consider the forest in a sustainable way. However, the knowledge of the importance of the photosynthetic process may be very relevant to discussions about sustainability.

Spirit is an attribute of everything in the universe. From an Indigenous perspective, it is a challenge to talk about living and nonliving things (biotic and abiotic), as a science curriculum does. Living versus nonliving is a Eurocentric dichotomy (Knudtson & Suzuki, 1992). According to Cajete (2006), Indigenous people experience and participate in the natural world not simply with their mind, but with their spirit, emotions, and body as well:

> Indeed, humans and the natural world interpenetrate one another at many levels, including the air we breathe, the carbon dioxide we contribute to the food we transform, and the chemical energy we transmute at every moment of our lives from birth to death. (p. 259)

It is common sense for many Indigenous people to see the physical and spiritual worlds as inextricably combined.

Nehiyaw (Plains Cree) scholar Willy Ermine (1995) characterizes Indigenous spirituality in metaphorical terms. He talked about an "inner space" (the metaphysical, intuitive, and spiritual world) and an "outer space" (the physical material world); both of which interact as one, according to monist and holistic presuppositions.

> Those who seek to understand the reality of existence and harmony with the environment by turning inward have a different, incorporeal knowledge paradigm that might be termed Indigenous epistemology. Indigenous people have the responsibility and the birthright to take and develop an epistemology congruent with holism. (p. 103) …
>
> Only by understanding the physical world can we understand the intricacies of the inner space. Conversely, it is only through journeys into the metaphysical that we can fully understand the natural world. (p. 107)

Ermine's powerful metaphor of an interacting inner spiritual and outer physical space is a recurrent theme in this chapter. A monist view of reality unifies these spiritual and physical experiences. Monism is a fundamental attribute that melds well with holism.

Holistic

Holism, sometimes spelled "wholism" by Indigenous people to convey the idea of something *whole*, rather than something holy, contrasts with Eurocentric reductionism. Holism reflects the assumption that parts of nature have meaning only in terms of their interrelationships with the whole of nature. Similarly, nature and culture exist as an integrated whole. Holism is also about balancing the mental, spiritual, emotional, and physical aspects of one's being. By splitting up and separating descriptions of a natural phenomenon into biology, chemistry, and physics, Eurocentric sciences (ES) disrupt the natural holistic way of experiencing the universe.

> No separation of science, art, religion, philosophy, or aesthetics exists in Indigenous thought; such categories do not exist. Thus, Eurocentric researchers may know the name of a herbal cure and understand how it is used, but without the ceremony and ritual songs, chants, prayers, and relationships, they cannot achieve the same effect. (Battiste & Henderson, 2000, p. 43)

Battiste and Henderson also point out that holism leads to "harmony as a dynamic and multidimensional balancing of interrelationships in [Indigenous people's] ecologies. Disturbing these interrelationships creates disharmony" (p. 43). According to Richard Atleo (2004), hereditary chief of the Nuu-chah-nulth Nation in British Columbia, an Indigenous view of reality goes beyond a Eurocentric concept of holism by incorporating spiritual and relational attributes into Indigenous holism. One Indigenous student defined holism as an action, "putting things that are apart back together again" (anonymous).

Relational

Elder Leroy Little Bear (2000) extended his explanation of a sacred world as follows: "If everything is animate [alive], everything has spirit and knowledge. If everything has spirit and knowledge, then all are like me. If all are like me, then all are my relations" (p. 78). As a constant reminder of their powerful relational existence, some Indigenous people end an Elder's prayer with an invocation that usually means, "All my relations." In Lakota this phrase is *Mitakuye Oyasin*, and in Plains Cree it is "*Kiyawow kichi nitotimak*." Carol Schaefer (2006) explains further:

> 'Mitakuye Oyasin' acknowledges that within each person exists the entire universe: all who have ever lived, all who are living now, and who are yet to be born, as well as nature—Our Mother Earth, the sun, moon, planets, and all the stars—all of the Sacred Universe since the beginning and until the end of time ... we are all cosmic beings. We come from the stars. (p. 146)

The expression "all my relations" proclaims a profound reality: As we make our way through life, we travel in a relational existence. Because all parts of life are interrelated, these relationships provide wholeness to existence.

> [A Navajo way of living in nature] may be viewed as the practice of an epistemology in which the mind embodies itself in a particular relationship with all other aspects of the world. For me as a Navajo, these other aspects are my relations. I have a duty toward them as they have a duty as a relative toward me. (Yazzie, 1996; as quoted in Cajete, 2000b, p. 64)

Indigenous ways of living in nature (IWLN) "tend to focus on relationships between knowledge, people, and all of creation (the natural world as well as the spiritual).... [IWLN requires] participating fully and responsibly in such relationships" (McGregor, 2002, p. 2). "Everything is one" (Atleo, 2004, p. 117) means a holistic network of spiritual relationships exists.

In Eurocentric thought, relationships are often viewed in terms of hierarchies, such as the Judeo-Christian tradition that places humans above animals, animals above plants, and so on. Significantly, Indigenous worldviews do not subscribe to this hierarchy (Cajete, 1999). Hence, everything in nature, including humans, enjoys equal status. For some Indigenous nations, humans have a lower status of importance than all other parts of Creation (Cajete, 2000b). Humility is a cherished value, rather than power and dominion over nature. To understand the natural world is to live in harmony with it, and not to dominate any part of it. Domination disturbs the balance and equality among relationships.

To acquire IWLN in an Indigenous context is to search for a balance among a web of relationships in a holistic sacred world. This is wisdom or more precisely, wisdom-in-action.

> Balance at the inner level [Ermine's inner space] is about maintaining a multidimensional equilibrium of physical, emotional, spiritual and intellectual development.... Balance at the outer level is about maintaining respectful interconnected, reciprocal and sustainable relationships beginning at the individual level embracing family, community, nation, and extending out toward the environment, plants, animals, and cosmos. (Michell, 2005, p. 40)

The Nîhîthewâk (Woodlands Cree) demonstrate their relationships with plants and animals through protocols and ceremonies. When outsiders learn some Indigenous knowledge about local plants and animals, they may or may not witness or be told the accompanying protocols and ceremonies. Some knowledge is so sacred it cannot be shared openly and publicly.

When everything is related and relationships require responsibilities, the whole of existence is made up of a web of interrelationships sustained by responsibilities. "As we experience the world, so we are also experienced by the world" (Cajete, 2006, p. 254). It is a reciprocal act. The act of observing, for example, includes a relationship between the observer and the observed. Trappers, for instance, know the behaviour, mannerisms, and habitat of the animals they hunt. Indigenous hunters personify their relationships with animals, which they may call their brothers, uncles, and grandfathers. In other words, Spirit, or according to some Elders, energy, flows through all Creation. How can the hunted be 'the other one' when 'the other one' is an extension of the hunter?

This Indigenous act of observing is the opposite of a typical scientist's "objective" observation, in which the observer is supposedly emotionally, morally, economically, socially, politically, and culturally detached. A relational, responsible world is a *personal world*, but a detached, objective world is *impersonal*. Students may resist learning science meaningfully because impersonal knowledge holds little or no value for them. The topic of vectors in physics provides a good example of unrelated, impersonal knowledge.

Interestingly, Barnhardt and Kawagley (2005) in Alaska point out how a Yupiaq act of observing is changing:

> As an elder completed the story of how he and his brother were taught the accrued knowledge associated with hunting caribou, he explained that in those days the relationship between the hunter and the hunted was much

more intimate than it is now. With the intervention of modern technology, the knowledge associated with that symbiotic relationship is slowly being eroded. (p. 9)

Knowledge gained through relationships established by repeated observations over time carries a responsibility of the Knowledge Keeper to both nature and the community. Traditionally, Indigenous people observed nature to learn how to live. For example, when a deer noticeably suffered from diarrhea, people might have observed what it ate as a remedy, because that information might be helpful for the same human problem. Tribe members who were gifted observers usually developed a rich understanding of nature, such as 'reading' the behaviour of certain birds to predict where a buffalo herd was located. These Knowledge Keepers were vitally important to and highly respected by their community, as they continue to be today.

The beginning of a respectful relationship with an Elder or Knowledge Keeper is expressed through proper protocol, such as offering a gift like tobacco or home-made jam (Saskatchewan Indian Cultural Centre, 2009). Elders or Knowledge Keepers can only pass their knowledge along to others who have formed an appropriate relationship with them. A Knowledge Keeper might share a physical, emotional, intellectual, or spiritual insight with a person, who could then live better through personal healing, political insight, or ceremonial knowledge, for example. Similarly, when one requests knowledge from a Nehiyaw Elder, one must first establish a relationship with the Elder, signified by his or her accepting a gift. Gifts are not a payment. Instead, acceptance of a gift is a way of acknowledging that a relationship has been formed. Everything is relational—all my relations!

A respectful relationship with natural forces such as thunder and lightning is similarly established. The sacred Thunderbird, for example, is revered as an explanation for thunder and lightning and as a protector of people who respect Thunderbird; a respect expressed by Ojibwa people, by their tobacco smoke rising into the stormy sky (CBC, 1995).

An Indigenous sense of relatedness has several positive outcomes. According to Four Arrows (2006), a person responsive to relatedness will (1) avoid dualistic thinking, (2) resist being a detached observer of nature, (3) emphasize cooperative engagement, not competition, and (4) live in harmony with nature rather than conquering it.

These descriptions fit Australian Indigenous peoples (Perso, 2003). For example, relationships permeate their mathematics. "Whereas non-Aboriginal people use number patterns based on counting and measurement, the patterns used by Aboriginal people [in Western Australia] are based on relationships between people [a genealogical pattern]" (p. 20). This sophisticated Indigenous mathematics was incomprehensible to British colonizers, and remains a challenge to Eurocentric thinking today. Although both sophisticated and useful to Australian Indigenous peoples, this genealogy-based mathematics system was not a priority in their culture because "Number does not carry the deterministic weight or the aura of objectivity and inevitability that it carries in non-Aboriginal Australia" (Watson & Chambers, 1989, p. 32). The scientific presupposition of quantification has little significance in many Indigenous cultures worldwide.

Mysterious

Indigenous ways of living in nature (IWLN) celebrate mystery and living in harmony with mystery in the inner and outer spaces of existence (Ermine, 1995). This fundamental characteristic contrasts with the scientific presupposition that nature is knowable through systematically analyzing nature and explaining it, thus elucidating its mystery. The opposite approach occurs with Indigenous Elders. Quite often they say they "know very little" and by doing so express the value of humility, which in this case means that no one person and no one society can ever come close to knowing all there is to know—"the Great Mystery," a metaphor for everything there is to know.

Another aspect of mystery in an Indigenous worldview is the idea of constant flux in nature. In contrast to a realist's or critical realist's assumption that reality is predictably the same and that observed changes in nature are predictable, IWLN view reality itself as being open to change. Some stories that convey IWLN teachings introduce a transformer (shape-shifter) or trickster who comes in many forms. Trickster is known by many names depending on the culture; examples include coyote, raven, hare, Nanabozho (Ojibwa), and Chikapash (James Bay Cree). Greg Cajete (2006) metaphorically associates trickster with chaos theory in Eurocentric sciences. Within the constancy of natural cycles, there always exists spontaneous and unpredictable flux.

> A perfect reflection of this cycle and transformation is the mythical figure Wisâkêchâk in our traditional [Woodlands] Cree stories. Wisâkêchâk reflects the notion of flux, change, continuity and interconnectedness as it transforms itself into various forms and crosses spiritual and physical boundaries, in order to teach people life lessons. (Michell, 2005, p. 37)

Cree journalist and playwright Doug Cuthand (2007) mentions several creation stories that involve the Plains Cree character Wesakechak,[20] but he cautions the reader to follow Cree protocol: "These stories are only to be told in the winter when the hard work is done and people need entertainment for the long winter nights" (p. 9).

One way to achieve harmony with the web of interrelationships for the purpose of survival is to coexist with the mysteries of nature. Harmony with nature is certainly not a romanticized notion for Indigenous peoples. The Woodlands Cree believe, 'We are the land, the land is a part of us.' Humans are only one strand within a sea of interconnected relationships.

Mystery also brings curiosity. Some Indigenous people consider their "laboratory" to be Mother Earth. Indigenous people are typically curious about observing phenomena and try to make sense of them in terms of IWLN. They focus on understanding meanings behind common patterns, rhythms, and cycles. Indigenous people also consider how other entities are connected to particular phenomena. Traditionally, a community's survival depended on people's intense curiosity and their collaborative success in knowing what the observations and phenomena meant. Curiosity encouraged by mystery leads to empirically based rational action for survival.

Dynamic

Indigenous ways of living in nature have evolved over time and continue to evolve in a way similar to scientific knowledge that changes due to new evidence or creative insights.

> It is important to realize that there is more to traditional knowledge than the repetition, from generation to generation, of a relatively fixed body of data—more than the gradual, unsystematic accumulation of new data over generations. In each generation, individuals make observations, compare their experiences with what they have been told by their teachers, conduct experiments to test the reliability of their knowledge, and exchange their findings with others. Everything that pertains to tradition, including cosmology and oral literature, evolves at the individual and community levels. Indeed, we suggest that the knowledge systems of Indigenous peoples are more self-consciously empirical

---

[20] The reader will have noticed that the same Cree word is spelled slightly differently. The spelling and pronunciation of Woodlands Cree and Plains Cree words can differ (as well for other major Cree groups). We follow the convention of using the author's spelling of a word. Variation in spelling can also be found among communities of the same Indigenous nation, and even among families within a community (e.g., Wîsahkecâhk, Wisakêtchak, and Wēsahkēcahk).

than those of Western scientific thought—especially at the individual level. Everyone must be a scientist to subsist by direct personal efforts as a hunter, fisher, forager, or farmer with minimal mechanical technology. (Battiste & Henderson, 2000, p. 45)

Interestingly, the word "scientist" in this quotation suggests that every culture has a science. Battiste and Henderson subscribe to a pluralist definition of science (see Chapter 3, the section "Clarification of *Science*").

Survival of Indigenous communities over millennia depends on a dynamic knowledge base.

> Indigenous people have traditionally acquired their knowledge through direct experience in the natural world. For them, the particulars come to be understood in relation to the whole, and the "laws" are continually tested in the context of everyday survival. (Barnhardt & Kawagley, 2005, p. 11)

For instance, the survival of First Nations in Canada today is challenged by loss of their traditional land to industrial and resource development, such as hydro dams and mining, and by the contamination of their water and traditional food resources (Simpson, 2004). Castellano (2000) predicts:

> The knowledge that will support their survival in the future will not be an artifact from the past. It will be a living fire, rekindled from surviving embers and fuelled with the materials of the twenty-first century. (p. 34)

Indigenous inner space (Ermine, 1995) guides people's reaction to spontaneous unanticipated flux in their outer space, and to their reassessment of that knowledge.

Indigenous ways of living in nature are not static; they are contemporary. They evolve dynamically with new observations, new technologies, new insights, and new spiritual messages (Kawagley, 1995). As nature changes, IWLN change. Non-contemporary IWLN that existed prior to contact with Europeans is often called "pre-contact," "historically traditional," or "ancient" knowledge.

Systematically Empirical

Systematic empiricism is an attribute common to IWLN and ES. It ensures a dynamic quality to IWLN, but it also serves Indigenous peoples in much richer ways. Elders systematically observe nature ("collect data" in ES) over many generations during which flux *naturally* occurs in their land. This flux is collaboratively discussed ("analyzed" in ES) in terms of changes (described as "variables" in ES) over long periods of time, along with the ensuing results. Elders accept natural events as they occur and never consider altering them. Scientists and engineers, on the other hand, cause change to occur *superficially* by exercising power over nature by controlling and manipulating nature—events they call experiments.

> Yupiaq scientific knowledge is based on thorough longitudinal studies and observations of the natural surroundings. Traditionally, knowledge was passed down from the elders to the youth through storytelling. Until recently, the Yupiaq language was not written down. Thus, all important knowledge was preserved by oral traditions that were crucial to survival. The preservation of the next generation depended on an efficient method of learning that which previous generations had already discovered (such as knowledge of seasonal and long-range weather patterns, salmon migration patterns, and knowledge about river ice and sea ice formation and movement). (Kawagley et al., 1998, p. 137)

These longitudinal observations of nature are like an engineer's experimental methodology that maximizes and minimizes variables. This approach contrasts with the control of variables or the accounting of variables used by scientists. "Behind these variables, however, there are patterns, such as prevailing winds or predictable cycles of weather phenomena that can be discerned through long observation (though climate change has rendered some of these patterns less predictable)" (Barnhardt & Kawagley, 2005, pp. 11–12). In a sense, then, this systematic, orally communicated, intergenerational observation of nature is *equivalent to* an engineering experiment—an Indigenous "experiment" perhaps. Moreover, this type of observation of nature has been integrated with Eurocentric ecology, for example, in traditional ecological knowledge (TEK) (McGregor, 2002; Nadasdy, 1999; Snively & Corsiglia, 2001; van Eijck & Roth, 2007) and in traditional ecological knowledge and wisdom (TEKW) (Menzies, 2006; Turner et al., 2000).

The meaning derived from Indigenous people's observations is connected to a holistic and sacred web of interrelationships sustained by responsibilities. "Through long observation [Indigenous peoples] have become specialists in understanding the interconnectedness and holism of our place in the universe" (Barnhardt & Kawagley, 2005, p. 12). Thus, unlike Eurocentric engineering, *Indigenous systematic empiricism enjoys holistic power*. Moreover, because Indigenous observations are monist in character, they relate systematically to both an outer physical space and an inner spiritual space well known to Elders within each community. Thus, *Indigenous systematic empiricism enjoys spiritual power*. Each generation's observations were passed along to, and incorporated into, subsequent generations' observations.

The community-based wisdom of the ancestors is a continuum of a profoundly rooted, yet ever-evolving, understanding of nature—holistic spiritual power.

This holistic spiritual power includes a wide range of observations such as dreams, visions, and subjective intuition (Castellano, 2000; Dyck, 1998; Michell, 2005). Many of these observations are collected systematically from teachings learned in vision quests, deep reflection, fasting, smudging, dreams, prayer, sweat lodges, and ceremonies (Saskatchewan Indian Cultural Centre, 2009). Some of this knowledge cannot be openly shared because it is sacred or because the knowledge holder does not believe a person is ready to receive it. Other teachings that emerge from these 'ancient methodologies' become part of the collective memory. Usually, all observations are analyzed collaboratively with wise Knowledge Keepers, who are often Elders, and then tested in the everyday world of personal experience. The source of these observations may be revelational from a Eurocentric point of view, but the process itself of making meaning is not. Schaefer (2006) talks about nature and spirituality:

> What needs to be developed [by humans who feel distant from nature] is a deeper, more personal sense of connection with the earth and our place in it. One way to create a deeper and more personal sense of connection is by holding rituals, ceremonies, and festivals. Ritual and ceremony are sophisticated social and spiritual technologies, refined by Indigenous peoples over many thousands of years, to celebrate and nurture the world of a particular place. Seasonal rituals and ceremonies speak to the whole community, which includes the plants, animals, and soil of a place and not just the human relationships. (p. 167)

When people live in a monist, sacred, holistic world situated in a specific place over long periods of time, their powers of observation expand. They become attuned to looking at multiple relationships that are not part of the consciousness of most scientists. Some scholars believe that this ability can help resolve our planetary crises (e.g., Cajete, 2000a; Davis, 2009; International Council of Science, 2002; Knudtson & Suzuki, 1992; Snively & Corsiglia, 2001). For instance, the expressions *listen to nature* and *taught by nature* are illustrated by Nakawē (Saulteaux) Elder Danny Musqua (1997) when he recounted events from his youth. Because of his small stature, his grandmother would send him into a beaver lodge for some of the "medicine" the beavers had meticulously collected and stored systematically in one of the lodge's small cavities. She had been taught which substances help different human ailments. This pharmaceutical knowledge may have been originally acquired through systematic study of beavers over long periods of time. It is in this sense that Indigenous people have become insightfully wise by "listening" to nature through empirical studies. In other words, Elders learn *from* birds, wind, and clouds, while scientists learn *about* them. "Everything in nature has something to teach humans" (Cajete, 2006, p. 254).

> When observed very carefully, animals reveal many secrets of living in balance and harmony. It is believed that animals have certain powers that can be used for personal, family, and community health and survival. (Michell, 2005, p. 40)

To observe nature means to inhabit nature, to engage with it in "a culturally conditioned 'tuning in' of the natural world.... None of this sensual participation with nature is 'supernatural' or 'extraordinary.' Rather, it is the result of an ancient and naturally conditioned response to nature" (Cajete, 2006, pp. 253–254). To engage with nature requires all one's senses and intuitive being in order to "enter into a heightened sense of awareness of the natural world" (p. 254).

Observations made by Indigenous people are usually qualitative, but Indigenous ingenuity with quantitative concepts is broadening the empiricism of Indigenous ways of living in nature.

> Over time, Native people have observed that the weather's dynamics are not unlike the mathematical characteristics of fractals, where patterns are reproduced within themselves and the parts of a part are part of another part that is a part of still another part, and so on. For indigenous people there is a recognition that many unseen forces are at play in the elements of the universe and that very little is naturally linear, or occurs in a two-dimensional grid or a three-dimensional cubic form. (Barnhardt & Kawagley, 2005, p. 12)

The acquisition of information for IWLN, therefore, is not restricted to a Euclidean world, as it is for most ES. Consequently, Indigenous systematic empiricism picks up on evidence that is figuratively and literally off the Eurocentric 'radar screen.' If one has sacred relationships and responsibilities with every part of the physical world, and if one is not constricted by the need to produce generalizable knowledge, then one's perceptions can be expanded—"a culturally conditioned 'tuning in' of the natural world" (Cajete, 2006, p. 253). The resulting diverse evidence gives advantage to place-based knowledge for the purpose of survival because one has more information to make a decision.

Based on Cyclical Time

A rectilinear concept of time is fundamental to most Eurocentric sciences. An alternative concept is cyclical time (Davis, 2009; Knudtson & Suzuki, 1992; Peat, 1994), a concept that harmonizes with the myriad of cycles observed in nature. Time has no beginning and no end. It reveals patterns, cycles, and rhythms. What goes around literally comes around.

> The idea of all things being in constant motion or flux leads to a holistic and cyclical view of the world. If everything is constantly moving and changing, then one has to look at the whole to begin to see patterns. For instance, the cosmic cycles are in constant motion, but they have regular patterns that result in recurrences such as the seasons of the year, the migration of the animals, renewal ceremonies, songs, and stories. Constant motion, as manifested in cyclical or repetitive patterns, emphasizes process as opposed to product. It results in a concept of time that is dynamic but without motion. Time is part of the constant flux but goes nowhere. Time just is. (Little Bear, 2000, p. 78)

Ermine (1995) reminds us that repetitive cycles in Indigenous outer physical space interact with cycles in inner spiritual space. Therefore, some spiritual ceremonies connect participants with past generations or with ancestral knowledge via their inner space. But these ceremonies are most effective when they occur in harmony with specific events in a cycle of the outer physical world. For example, many Blackfoot and Lakota communities on the Canadian plains perform the Sundance ceremony at summer solstice. On the west coast of Canada, winter dancing by Haida and Nuu-chah-nulth communities takes place at the winter solstice. These solstice occurrences maximize the power of the earth and heavens in Indigenous ceremonies that penetrate an Indigenous deep space of time. A Eurocentric worldview might view connecting with past generations as time travel into the past, but that perspective is filtered through the lens of rectilinear time. Seen through the lens of cyclical time, it is not time travel but a natural relationship in the web of interrelationships. Some people of the Dëne Nation refer to these types of movements around time as dream quests, allowing people to see events of both the past and the future.

The human pulse of cyclical time is illustrated by Nakawē Elder Danny Musqua (quoted in Knight, 2001):

> We have a beautiful tradition and a holistic view of the universe that makes us who we are. In our circle, we need the old and the young, the old to teach and the young to keep the tradition alive. Nothing really dies out in a circle, things might get old and wear away but they renew again, generation after generation. That is what the circle is about. (p. 5)

In a way similar to Eurocentric time, Indigenous time is also the space between two events. However, this space is much more than a numerical, *physical*, measured space. The space of time between events also includes *emotional, mental,* and *spiritual* dimensions. All four dimensions are part of the Indigenous concept of time.

When beginning an important event or meeting, Elders talk about 'when the time is right.' That is, an event should begin when the leaders sense that people are physically, emotionally, spiritually, and intellectually in synchrony with the event. This is why a smudging ceremony is often used to begin a council meeting. The ceremony focuses people's readiness to participate and helps a leader decide when to begin. A decision is not reached by looking at one's watch—the icon of rectilinear time. And so goes the Indigenous adage: 'White people have watches; Indians have time.'

Valid

Michell (2005) spoke for Indigenous peoples worldwide when he wrote, "Woodlands Cree cultural knowledge needs no validation from Euro-Western knowledge systems" (p. 37). A Dakota Elder expressed it this way:

> First Nations knowledge ways are a product of a valid human experience in a relatively similar environment that has been developed over a long period of time; and for this reason, it is valid in its own right. Validity has to do with the fact that First Nations people are still here today. We have survived using our own knowledge systems. (Michell et al., 2008, p. 76)

As discussed in Chapter 4 (the section "Eurocentric Sciences Have Predictive Validity"), the validity of ES is restricted to predictive validity or predicting *how* the physical universe works. The validity of IWLN is based on content validity, which deals with the issue of *what* the universe *is*.

> Indigenous peoples throughout the world have sustained their unique worldviews and associated knowledge systems for millennia, even while undergoing major social upheavals as a result of transformative forces beyond their control. Many of the core values, beliefs, and practices associated with those worldviews have survived and are beginning to be recognized as being just as valid for today's generations as they were for generations past. The depth of Indigenous knowledge rooted in the long inhabitation of a particular place offers lessons that can benefit everyone, from educator to scientist, as we search for a more satisfying and sustainable way to live on this planet. (Barnhardt & Kawagley, 2005, p. 9, emphasis added)

The "depth of Indigenous knowledge systems" mentioned by Barnhardt and Kawagley hints at content validity. Any knowledge system that has succeeded for such a long time must have content validity. The evidence is time plus survival. Indigenous content validity holds promise to be highly relevant to resolving and avoiding environmental crises.

Battiste and Henderson (2000, p. 121) likened the validity of Indigenous ways of living in nature to intelligible essences, which are the source of an idea's or an event's distinctive identity—what it *is*. Our intellect gives birth to abstract concepts when it apprehends the intelligible essences of external realties. Ideas of what is valid and invalid have their distinctive identities tied up in intelligible essences abstracted by the intellect. Indigenous content validity goes a step further by also recognizing external realties as being both emotional and spiritual.

Intelligible essences are rejected by Eurocentric sciences. But, according to Aristotle, if we deny the validity of intelligible essences, we are denying the fidelity to a true world. One alternative to 'a true world' is a 'represented world;' a world made up of thought processes and metaphors that *represent* the material world and which have a high degree of success in predicting what will occur in the material world. We are speaking of course of ES (see Chapter 4, the section "Reality Is Reproduced or Represented by Scientific Knowledge"). The content validity of IWLN and predictive validity of ES complement our understanding of nature. They coexist. Both rely on logical reasoning. Both are true, in a pluralist sense.

Although the power to predict is essential for Indigenous peoples to obtain subsistence from nature and to survive, the short-term immediate predictive power of IWLN does not compare favourably with the immediate predictive power of many Eurocentric sciences. Instead, the validity of IWLN lies elsewhere—in its content validity.

Barnhardt and Kawagley (2005) note the following two points: First, the 450 years since Copernicus published his model of the heavens is a relatively short period of time compared with the tens of thousands of years associated with IWLN. Second, the ensuing disastrous impact of material progress on planet Earth during those 450 years, made possible by the participation of Eurocentric sciences and technologies, challenges ES's content validity in terms of humanity's survival. This sentiment is echoed by many other authors, such as Cajete, 2000a; Davis, 2009; International Council of Science, 2002; Knudtson & Suzuki, 1992; and Snively & Corsiglia, 2001).

Battiste and Henderson (2000) argue that for anyone "engaged in a lifelong personal search for ecological understanding, the standard of truth in

Indigenous knowledge systems is personal experience" (p. 45). Success in the everyday world of personal experience is a much different criterion from the criteria used by scientists during their consensus-making processes to determine "scientific truth" (see Chapter 4, the section "A More Realistic View of Eurocentric Sciences"). Truth for Indigenous peoples can be gained through "the infinite moments of both chaos and order. This is a precept of Native science, for truth is not a fixed point, but rather an ever-evolving point of balance, perpetually created and perpetually new" (Cajete, 2006, p. 253).

Elders teach us that:

> We are not so much meant to discover the one true picture of reality, but rather we are meant to construct the fullest and clearest picture of the situation we can, by integrating our best collective knowledge. The more viewpoints and ideas included, the more complete and meaningful the picture will be. Knowledge embedded in context and interpreted from a network of perspectives has the opportunity to be rich in metaphors. It is not only the perspective of the people engaged in the dialogue whose views must be taken into consideration, but ideas are always examined against views of the ancestors embedded in people's memory and in the stories, songs, and dances. Equally, the viewpoints include future generations and how current decisions will affect them and their world. (Snively & Williams, 2008, p. 125)

This pluralist richness of truth presents a particular challenge to the doctrines of realism and positivism (see Chapter 4). The Elders' idea also means that when science teachers present different viewpoints on knowing nature, students are able to construct a fuller and clearer picture of nature (Eurocentric and Indigenous meanings of nature; see Chapter 5, the section "Clarification of *Knowledge* and *Nature*"). Their picture will most likely correspond to their own worldviews and cultural self-identities. Accordingly, school science becomes more meaningful and encourages students to learn the best of both knowledge systems.

Rational

Mathematics, undeniably a rational way of knowing, is built upon a set of axioms. Axioms are neither true nor false, they are simply assumed. Axioms are a characteristic of rational knowledge systems. From them, a knowledge system is developed. As described in Chapter 4 (the section "The Material World Is Governed by Quantification"), if we change a single axiom in Euclidian geometry, we get a different geometry system such as Riemann geometry that allows "parallel" lines, such as lines of longitude, to meet.

Chapter 4 identified key axiom-like ideas—that is, culture-based presuppositions—found in Eurocentric ways of knowing nature. If one uses these axiom-like ideas consistently, along with applicable empirical data and other scientific ideas, a person would be engaged in a rational system of reasoning.

Similarly, Indigenous ways of living in nature are built upon axiom-like ideas, several of which are described in the present chapter. Elders and Knowledge Keepers demonstrate a consistent use of logical reasoning that flows from their axiom-like ideas, along with systematically empirical observations. Therefore, an Indigenous way of knowing nature is a rational system of reasoning. Like that of Eurocentric sciences, Indigenous rationality is culture-based because one's culture determines what axioms are assumed. For example, in many ES, it is rational to believe that the material world is a mathematical world, and so scientists emphasize measurement. However, in IWLN it is rational to believe that the material/non-material world is a spiritual world, and so Elders emphasize intuition.

Understandably, most Elders or scientists have faith or unquestioned acceptance in their own group's axioms, and cherish their own group's rationality. Conflict can arise when someone cherishes his or her own cultural rationality to the extent of denying that of a different culture. For example, the faith scientists hold in their own rationality causes many of them to think that IWLN are based on superstition. But, at the same time, faith in Indigenous rationality causes some Indigenous people to think that ES are equally based on superstition. Lakota Elder Deloria (1992) stated:

> The present posture of most Western scientists is to deny any sense of purpose and direction to the world around us, believing that to do so would be to introduce mysticism and superstition. Yet *what could be more superstitious* than to believe that the world in which we live and where we have our most intimate personal experiences is not really trustworthy, and that another mathematical world exists that represents a true reality? (p. 40, emphasis added)

Thinking that someone is rational or superstitious may depend upon which culture-based knowledge system one adheres to. A resolution to this conflict is possible when we are open-minded about allowing diverse knowledge systems to complement each other. This understanding requires science teachers to 'think outside the box' and to allow the exchange of multiple perspectives in their classrooms. Such approaches occur naturally when science teachers construct bridges between their scientific knowledge system and an Indigenous knowledge system (Belczewski, 2009).

Spiritual

As evident in the quotations throughout this chapter, spirituality is pervasive in Indigenous ways of knowing nature. Chickasaw scholar Eber Hampton stated: "The first standard of Indian education is *spirituality*. At its centre is respect for the spiritual relationships that exist between all things" (Hampton, 1995, p. 19). A Lakota ceremonialist commented:

> This is not a scientific or technologic world. The world is first a world of spirituality. We must all come back to that spirituality. Then, after we have understood the role of spirituality in the world, maybe we can see what science and technology have to say. (quoted in Simonelli, 1994, p. 11)

Ojibwa author Richard Wagamese (2008) wrote:

> Everything is energy. This is what our [traditional] teachers say. Great Spirit is the feeling of that energy expressed in all things, radiating everywhere around us.... We are all one being. We are all one soul, and we need each other. That is spiritual. That is truth. That is Indian. (p. 209)

Religious people certainly possess their own spirituality—"seeking understanding beyond the physical world" (Saskatchewan Indian Cultural Centre, 2009, p. 14)—but this spirituality likely differs fundamentally from Indigenous spirituality—"a relationship with all of creation" (p. 14). Spirituality is a way of life for many Indigenous cultures around the world. This nature-based way of life "gives sacred significance to all things" (Four Arrows, 2006, p. 279). Because Spirit flows through all of existence, the spiritual component of a human being cannot be separated from the day-to-day life of students and their communities.

Spirituality assumes monism—the assumption that the physical and metaphysical worlds are one domain, not two separate ones as in Cartesian dualism. This was Ermine's (1995) idea when he described the intermingling of our material physical "outer space" and our metaphysical, intuitive, spiritual "inner space." This intermingling is central to Indigenous rationality. Indigenous "spirituality is the sense of relationship or connection with all parts of Creation, each of which has 'Spirit'.... Many responsibilities accompany this relationship. Together, the responsibilities and the relationship make up spirituality" (Coalition for the Advancement of Aboriginal Studies, 2002, p. 283). Furthermore, "Death is understood as a metamorphosis, wherein the spirit of the deceased does not disappear, but becomes part of the animating and creative forces of nature" (Cajete, 2006, pp. 254–255).

Elder Andy Blackwater (2009) of the Blackfoot Nation also speaks of Spirit flowing through everything in Creation, so that everything is alive with Spirit. It is rational, therefore, to communicate with Spirit in an animal or plant. If you hurt Spirit in other beings in nature, you hurt the very essence of your own spirit; much like poisoning the water systems is identical to poisoning humankind. It is rational to think of Spirit flowing through all humans, regardless of their religion, ancestry, or beliefs. In addition, Spirit carries energy that can be focused on physical parts of Creation through words, such as prayers or the lyrics of songs, which have the potential to affect physical phenomena because the non-material world intermingles with the material world.

The keepers of spiritual understandings are most often Elders. They teach their community's spirituality to people who are experiencing the land, where spiritual awakening and awareness can occur. In-depth teaching of Indigenous spirituality is not appropriate in science classrooms (Michell et al., 2008). However, *respecting* people's spirituality, whether Indigenous or non-Indigenous, is a core value for all teachers. Elders often say, "There are many paths to the same source."

Showing respect for Indigenous spirituality comes in many forms (Saskatchewan Indian Cultural Centre, 2009). For example, a teacher could find out which nature stories are appropriate to repeat and what protocols should be followed when reading or telling them. A teacher could also ask in a private conversation what topics in a local IWLN may have particular spiritual significance to students. A person expresses respect by simply asking such questions. If these topics ever arise in the classroom, teachers could simply acknowledge them as having sacred importance. However, teachers from outside that community, or who are not Indigenous, should seek permission and guidance from the community on how to proceed.

Students can become aware of powerful symbolic metaphors, such as Thunderbird, that connect to Indigenous spirituality without discussing or accepting that spirituality. The Medicine Wheel, also known as the Circle of Life by some Indigenous people, is a symbol central to Indigenous spirituality. The number four is considered a spiritual number. The four directions of the Medicine Wheel—east, south, west, and north—unite many natural and social phenomena in ways that ultimately relate to Indigenous spirituality. For instance, in the Nehiyaw (Plains Cree) culture, the four winds (*Awpun, Sawin, Nepawanuk,* and *Kewatin*) have significant stories associated with living in the natural world in particular ways, which are summarized in Cuthand's (2007) book *Askiwina*. A cross-cultural science teaching unit, "Nature's Hidden Gifts" in *Rekindling Traditions* was developed on the basis of the Circle of Life (Aikenhead, 2000). The unit exemplifies a respectful understanding of Indigenous spirituality without teaching spirituality in-depth. Science teachers need to distinguish between students' *understanding* classroom content and students *believing* that content by making it part of their identity (Cobern, 1996). This is a powerful distinction for science educators and administrators to make (see Chapter 7, the section "Deborah's Story").

From an Indigenous perspective, respect for Indigenous peoples' spiritual relationship to the land has political ramifications. One's duty to ask permission to use Indigenous lands extends into the realm of Indigenous knowledge systems because the two are inseparable. "Respect for Native spirituality and the nature-wisdom embedded within it is inseparable from respect for the dignity, human rights, and legitimate land claims of all Native peoples" (Knudtson & Suzuki, 1992, p. 18).

The Concept of Knowledge Revisited

Chapter 5 earlier described coming to know Indigenous knowledge as a journey toward wisdom-in-action. Now, having discussed the fundamental attributes of Indigenous ways of living in nature (IWLN), this chapter revisits an Indigenous concept of knowledge through the writings of four Indigenous scholars in North America—from the Mi'kmaw, Chickasaw, Nîhîthewâk, and Pueblo Nations. Also included is an Inuit perspective published by the Nunavut Government in Canada.

In the context of clarifying Indigenous knowledge, Marie Battiste (Mi'kmaq) and Sakej Henderson (Chickasaw) (2000) summarized the following points about IWLN:

> Indigenous peoples regard all products of the human mind and heart as interrelated within Indigenous knowledge. They assert that all knowledge flows from the same source: the relationships between a global flux that needs to be renewed, the people's kinship with other living creatures that share the land, and the people's kinship with the spirit world.... (p. 41)
>
> Indigenous ways of knowing share the following structure: (1) knowledge of and belief in unseen powers in the ecosystem; (2) knowledge that all things in the ecosystem are dependent on each other; (3) knowledge that reality is structured according to most of the linguistic concepts by which Indigenous people describe it; (4) knowledge that personal relationships reinforce the bond

between persons, communities, and ecosystems; (5) knowledge that sacred traditions and persons who know these traditions are responsible for teaching "morals" and "ethics" to practitioners who are then given responsibility for this specialized knowledge and its dissemination; and (6) knowledge that an extended kinship passes on teachings and social practices from generation to generation. (p. 42)

Similarly, Herman Michell (2005) (Nîhîthewâk) wrote about different types of knowledge:

From an Indigenous worldview, all living things are endowed with a conscious Spirit. From this understanding Woodlands Cree knowledge is manifested in different forms, some of which is practical and learned through day-to-day activities that revolve around survival. Our people also possess empirical knowledge that is learned from careful observations of the natural world over extended periods of time. There are other types of knowledge that link with ceremonial ways that need to be handled with extreme sensitivity. The "revelatory" knowledge is often assessed through elders' guidance, consultation, and preparation; using proper protocols, including dreaming and visioning. Certain knowledge is given to people when they are ready to receive it. (p. 38)

Greg Cajete (1999), a Tewa member of the Santa Clara Pueblo Nation, points out some features of IWLN that subtly parallel ES and others that do not:

Indigenous science is internally consistent and self-validating. Its definition is based on its own merits, conceptual framework, practice and orientation. It is a disciplined process of coming to understanding and knowing. It has its own supporting metaphysics about the nature of reality. It deals with systems of relationship. It is concerned with the energies and processes within the universe. It provides its own basic schema and basis for action. It is fully integrated into the whole of life and being, which means that it can not be separated into discrete disciplinary departments. (p. 84)

In a later publication Cajete (2006) contrasts this view with ES:

The Western science view and method for exploring the world starts with a detached "objective" view to create a factual blueprint, a map of the world. Yet, that blueprint is not the world. In its very design and methodology, Western science estranges direct human experience in favor of a detached view. It should be no surprise that the knowledge it produces requires extensive recontextualizing within the lived experience in modern society. This methodological estrangement, while producing amazing technology, also threatens the very modern life-world that supports it. (p. 257)

The Department of Human Resources (2005), Government of Nunavut in Canada, collaborated with the Inuit Nation to describe traditional Inuit knowledge—*Inuit Qaujimajatuqangit*. It is characterized in terms of the following six guiding principles:

1. Concept of serving (*Pijitsirarniq*): "Each person has a contribution to make and is a valued contributor to his/her community"
2. Consensus—decision making (*Aajiiqatigiingniq*): "Being able to think and act collaboratively, to assist with the development of shared understandings, to resolve conflict in consensus-building ways, and to consult respecting various perspectives and worldviews"
3. Concept of skills and knowledge acquisition (*Pilimmaksarniq*): "Building personal capacity in Inuit ways of knowing and doing.... Demonstrating empowerment to lead a successful and productive life, that is respectful of all"
4. Concept of being resourceful to solve problems (*Qanuqtuurungnarniq*): "Innovative and creative use of resources and demonstrating adaptability and flexibility in response to a rapidly changing world"
5. Concept of working together for a common purpose (*Piliriqatigiingniq*): "Stresses the importance of the group over the individual"
6. Concept of environmental stewardship (*Avatimik Kamattiarniq*): "Stresses the key relationship Inuit have with their environment and with the world in which they live"

Inuit Qaujimajatuqangit is also illustrated in Inuit traditional stories published by the Nunavut Bilingual Education Society (2004, 2006).

By revisiting the Indigenous concept of knowledge, one realizes that its meaning is not accurately represented by the expression "Indigenous knowledge." More accuracy and clarity comes from the expression "Indigenous ways of living in nature." However, the general public is not familiar with such an expression. Consequently, Indigenous scholars continue to talk about "Indigenous knowledge" in order to communicate with their audience, even though the Eurocentric concept of knowledge is by and large incommensurate with IWLN.

Conclusion

The more one learns about Indigenous ways of living in nature (IWLN), the more one appreciates the ideas that normally get lost in translation from one knowledge system to another. This chapter incorporated these lost ideas into its discussions about Indigenous ways of knowing nature. In doing so, the chapter moved away from the old colonial way of speaking in which dichotomies privilege a Eurocentric way of thinking, and toward a new *postcolonial* way of speaking, The term 'postcolonial' does not mean that colonialism has ended. Instead, it means that colonialism is explicitly recognized and efforts are being made to diminish and extinguish its power, a process called *decolonization* (Belczewski, 2009; Chinn, 2007; Saskatchewan Indian Cultural Centre, 2009).

Understanding IWLN is a moral act as well as an intellectual achievement for science teachers because this understanding is a step toward the following:

- bridging one's scientific worldview and Indigenous worldviews for the benefit of both Indigenous and non-Indigenous students; and
- preparing to implement an enhanced science curriculum, and negotiating an inclusive meaning for school science in one's community.

By understanding IWLN, one also gains insight into the cultural ways of industrial countries. In Euro-Canadian culture, for instance, modern technology has greatly reduced people's direct reliance on, and connection to nature, virtually eliminating their intuitive participation in the natural world. When people do not directly benefit from intuitive participation, they face a paradox. On the one hand, they are distanced and alienated from nature by technology; on the other, they feel technologically in control of nature for improving their material quality of life. Environmental degradation often ensues from this paradox. Wagamese (2008) expresses the idea differently by capturing a positive perspective inherent in his Ojibwa culture:

> The science of the earth [IWLN] is a different creature from the science of numbers and theorems. It's a discipline of co-existence. It's the knowledge and acceptance of the mystery that surrounds us—and the awareness that allowing it to remain a mystery, celebrating it rather than trying to unravel it, engenders humility and a keen sense of the spiritual. (p. 126)

Most Indigenous people acknowledge that returning to their former hunter-gathering and agricultural ways is not practical. But they want to sustain their ancestral wisdom-in-action that connects them experientially, intuitively, and spiritually with Mother Earth (Battiste, 2002; Cajete, 2006; Michell et al., 2008). Despite the ravages of colonization, many Indigenous people have retained a core worldview and philosophy of life that can be drawn upon to rethink how humans can live out their lives in relationships with all of Creation. Henry Lickers, environmental biologist of the Seneca Nation, eloquently summarized his people's worldview and philosophy:

> First Nations people view themselves not as custodians, stewards or having dominion over the Earth, but as an integrated part in the family of the Earth. The Earth is my mother and the animals, plants and minerals are my brothers and sisters. (Canadian Council on Learning, 2007b, p. 2)

CHAPTER 7

Comparing the Two Ways of Knowing Nature

Chapters 3 to 6 described two different culture-based ways of knowing nature, Eurocentric sciences (ES) and Indigenous ways of living in nature (IWLN). Although there is great diversity within IWLN, Indigenous cultures share common features that provide a valuable starting base for teaching IWLN in school science. For example, Indigenous peoples share a strong connection to the land because of their hunting, fishing, and gathering activities, as well as producing products from their place-based land and marine resources. From this intimate relationship with the land and their trading practices between nations, Indigenous peoples developed some common features of their worldviews in which humans are interdependent with the natural world.

This book takes the pluralist position that there are multiple ways of knowing nature. ES are but one strand within a complex network of traditions worldwide, many of which originate among Indigenous peoples. This position contrasts with a universalist perspective, which allows only one way of legitimately knowing nature.

We have learned that IWLN and ES have *common* features and have *different but complementary* ways of dealing with nature. We do not say they are equal but we know they coexist. "Different but complementary" is similar to the coexistence of yin and yang in Eastern philosophies.

Although Indigenous knowledge systems need no validation from ES, some scientists have found tremendous value in coming to know IWLN for resolving some of humankind's shared planetary problems (Snively & Corsiglia, 2001). Both knowledge systems deserve recognition, respect, and understanding.

This chapter highlights the comparisons of key characteristics of Eurocentric sciences and of Indigenous ways of thinking, reflecting, living, and being. This is accomplished first, by listening to the experiences of an Indigenous science student; second, by reminding ourselves how to avoid pitfalls when comparing cultures; and third, by summarizing ideas discussed in previous chapters.

Deborah's Story

We begin this chapter with a vignette of the successful experiences of a science student dealing with differences between ES and her own Indigenous worldview as a Diné (Navajo). This vignette draws upon extensive research undertaken by Carol Brandt (2008) at a detailed level of observation and analysis with Deborah, a university BSc biology major. Because the discipline of biology has strong links to the natural world, one might think that any person of Indigenous heritage would have a natural inclination toward this field and succeed with high marks. However, the following story illustrates some of the complex challenges and learning difficulties faced by Indigenous students as they attempt to cross back and forth between a Eurocentric science and their Indigenous way of thinking, reflecting, and being.

Unfortunately, almost no in-depth research about Indigenous students in schools has been reported. We hope that many vignettes will soon be written by teachers who implement cross-cultural science instruction with Indigenous students, and who conduct action research with their students to give voice to students' experiences. In the meantime, Deborah's story, the only one publicly available at this time, can help us understand what many Indigenous students experience in school science.

Deborah had always loved science in school. When attending high school, she decided to become a physician to help her people in New Mexico improve their standard of health and ensure their cultural survival. Deborah's challenges in undergraduate biology classes were many. Because her first language was Diné, she struggled with the way people normally communicate in ES. The technically precise jargon used in ES is esoteric and foreign to many students (Hodson, 2009, Ch. 8), particularly Indigenous ones (Kawasaki, 2002). When confronted with difficult concepts in ES, she tried putting them into the familiar context of her Diné thinking and speaking. Invariably, she had trouble doing so because the worldviews associated with each language were so different. In frustration, she told one instructor, "A lot of what you are saying I can't understand because it's not in my world" (Brandt, 2008, p. 837).

It was not until her fourth year at university when Deborah finally realized that she needed to distinguish between *understanding* a scientific idea and *believing* it. As described below, this conceptual distinction lifted a tremendous weight off her shoulders. No instructor ever helped her resolve the cultural clashes with her Diné worldview. Nor did any instructor show her how to challenge or question scientific ideas. "They give this evidence and there's no argument.... [they assume] it must be true," explained Deborah (p. 835). She often referred to ES as a process of "belief" rather than a process of evaluating scientific data. Instructors would tell her just to accept it. Their attitude made Deborah feel like an outsider who was expected to devalue, or even abandon her own identity and take on one similar to the professor's universalist ideology.

Deborah grew up understanding that animals, plants, and other life forms were an extension of her self. Like many Indigenous peoples around the world, she understood that everything is spiritually imbued. In Diné culture, there are proper protocols, prayers, and ceremonies involved when taking the life of an animal. Offerings are made in order to restore balance. The Diné foundational concept of harmony conflicted with ES when she was expected to dissect frogs and work with cadavers with emotional detachment—"invading others" as Deborah called it. She concluded that this invasion "disrupts the harmony within ourselves and then we are not the person we should be" (p. 836).

Deborah also held strong beliefs in the Diné Creation Story and the origins of her people. This grand narrative conflicted with scientific evolution and natural selection in Eurocentric biology. It was some time before Deborah was able to accept the existence of both stories without having to sacrifice one over the other.

Although the molecular biology of viruses was difficult to translate into Diné, she could actually see infected cells. They were concrete, unlike the concept of evolution that could not be experienced directly. Moreover, ideas about viruses did not marginalize her cultural identity.

Deborah had the good fortune to be hired by a molecular biology research laboratory during the summer between her third and fourth years. Unlike her experience in academic courses where discussion about her culture clashes was unacceptable, she found her laboratory co-workers and mentors to be much more open. They helped her communicate in the culture of microbiology. She could ask questions, challenge their ideas, and make connections between the biological knowledge she had previously memorized and specific lab procedures (knowledge-in-action). She formed relationships with these people, unlike her experience with instructors. She began to understand ES in a deeper, more meaningful way, which in turn led her to adopt strategies for completing a BSc program with her Diné identity intact.

"What I'm learning from my non-Navajo world will help my people health wise. But what I've been brought up with, it's there. I'll always believe my creation story" (p. 836). Deborah finally realized that Diné tradition and Western medicine could be *complementary* and promote harmony at the same time. Her newly gained point of view allowed her to risk participating in Eurocentric biology without having to change her beliefs. Instead, she could temporarily suspend any paradoxes with her Diné world. The two cultural ways of knowing nature began to coexist for Deborah, thus lessening the risk of culture clashes.

Deborah achieved high grades during her fourth year at university because she had learned how to succeed in an educational system without losing her cultural identity. Her instructors, however, needed to develop a balance between teaching ES and recognizing Indigenous worldviews, a balance that emphasizes academic achievement and students' cultural identities (Brandt, 2008).

Deborah's story highlights the importance of fostering supportive peer relationships and establishing classroom environments that honour and respect Indigenous perspectives. Indigenous students in general face many systemic barriers that keep them from entering or succeeding in postsecondary science, technology, and engineering programs. In particular, Deborah's story speaks to the challenges caused by worldview clashes, and the role of language in those clashes.

The languages of Deborah and her instructors were worlds apart. Instructors need to consider allowing longer response times when asking questions, so that an Indigenous student has time to make the necessary cognitive translations before answering. For example, teachers could pose key questions at the beginning of a unit, or a week, or a lesson—questions that give students guidance in identifying the main points of the instruction. This would allow time for students to reflect on what they know, to sort out their thoughts, and to respond in a more holistic manner. Some of these questions might appear on a unit examination. At the same time, Indigenous students must learn the languages and vocabulary appropriate for different scientific contexts in order to function professionally in collaboration with mainstream society.

Hidden Pitfalls to Avoid When Comparing Two Cultures

Comparisons of IWLN and ES can cause misunderstandings. Some hidden pitfalls are summarized here for the purpose of highlighting strategies for science teachers' clear thinking and sensitivity to culture. These pitfalls include stereotyping, language, and different versions of Eurocentric science.

Stereotyping

The first pitfall concerns the vulnerability of IWLN and ES to stereotyping, especially when their similarities and differences are discussed. Stereotyping is reduced when we are particularly conscious of each group's diversity, and if we treat general statements as being *indicative* of a group, rather than *prescriptive* of a group. The diversities found within IWLN and ES have been detailed earlier. In this chapter, comparisons of IWLN and ES are offered in the context of those diversities with the clear intent *to indicate* prominent features in each group.

One needs to be aware that questionable early research and writings about Indigenous people have contributed very much to the way in which Indigenous people are perceived and portrayed today. Stereotypes, racist beliefs, and unwarranted generalizations continue to be evident in such Eurocentric institutions as universities and schools.

Language

Translation has the potential to clarify our understanding of another culture. But words or sentences are often difficult to translate with precision, and consequently misunderstandings arise. Thus, language harbours several pitfalls when comparing cultures.

We need to know that languages such as Cree and English or French (the other colonizing language in Canada) are not parallel, which means that a word or idea expressed in Cree likely does not have the exact corresponding word or idea in English, and vice versa (Kawasaki, 2002).

Dictionary translations lure us into a false belief that words have equivalent meanings or identical concepts. Much is distorted or lost in translation as a result, especially between European and Indigenous languages. Listeners or readers could easily misunderstand an Indigenous idea expressed in English. Marie Battiste (2002, p. 2) reminds us, "There are limits to how far [an Indigenous knowledge system] can be comprehended from a Eurocentric point of view." That is one reason Elders offer sacred prayers only in their native tongue. The best defence against the pitfall of what is 'lost in translation' is to be vigilantly open-minded about its possibility.

An exception to the problem of translation appears in some Indigenous words that came into existence after contact with Europeans. Some words have clear conceptual translations. For example, "computer" translated into Inuktitut means "a machine that thinks."

We need to be sensitive to the reality that language is the glue that holds a culture together. It is a repository of cultural understanding that sustains a collective worldview. Single words can contain complex understandings of the natural world. A humorous example is the Woodlands Cree word *makiti*, which means either big heart or a huge posterior. Although Indigenous languages are the best means of communicating Indigenous knowledge, this is not possible in many schools. However, the few Indigenous phrases and vocabulary a teacher might learn would go a long way toward making Indigenous students feel more comfortable in class.

Language is also a currency of prestige and power. In general, the language of ES carries much greater authority than the language of any Indigenous group. Thus, a power imbalance exists between IWLN and ES in terms of each culture's language. One way to reduce this imbalance is to acknowledge its existence. Expressions such as "Indigenous science" or "Native science" promote greater equity with the expression "Eurocentric science."

Our intended sensitivity toward worldviews different from our own can be very difficult to achieve in the context of written and oral language. When bilingual Elders talk to us in English because we do not speak their language, their English words can unintentionally force them into a Eurocentric point of view. As a result, we may misunderstand their message. In the same way, comparisons between IWLN and ES become unavoidably tainted with the brush of Eurocentric thought because we are communicating in English. Chapter 5, for example, showed how the word "knowledge," steeped in Eurocentric thought, has a very different meaning in the culture of ES compared to Indigenous cultures (Cajete, 2006, p. 250). Our solution to minimize this difficulty was to replace "knowledge" with "ways of living," so the Eurocentric idea of knowledge did not distort a reader's awareness and understanding of an Indigenous perspective.

Our comparisons of IWLN and ES in this book are inescapably lodged within the English language. We acknowledge that an Anglo perspective dominates our discussion. However, the problem can be reduced over time by engaging in dialogue. Sensitivity toward the Anglo-language-monopoly problem is sharpened by asking ourselves: Whose language is being spoken? Do we acknowledge that one group is thereby privileged by the use of their language?

When making a presentation in English or French, many Indigenous speakers subtly acknowledge Anglo or Franco language monopoly by giving the introduction to their talk in their mother tongue. Another approach could be that Anglophone or Francophone teachers begin their presentation to Indigenous people by expressing a greeting in the people's language. For example, the teacher could say in Plains Cree, *Tān'si. Namōya mānitaw*, which means "Hello, how are you? and then shift into English/French with good-hearted apologies for not being able to continue in the Plains Cree language. That makes people aware that the speaker is conscious of the language-monopoly problem.

Different Versions of Eurocentric Sciences

In comparisons of IWLN and ES, a third pitfall arises concerning which version of ES will be presented. Will it be the conventional, outdated, idealized, positivist version normally found in mainstream society? Or will it be a revised version based on the scholarship presented in Chapters 3 and 4? The revised version reflects post-positivist developments in science education variously identified as science-technology-society-environment (STSE), the nature of science (NOS), and socio-scientific issues (SSI). The positivist version of ES denies the validity of IWLN (see Chapter 4, the section "Empirical Data Speak for Themselves: Positivism"). The revised version embraces a pluralist understanding of science that will recognize IWLN as foundational in school science.

Summary

The pitfalls described above have the capacity to undermine comparisons of Indigenous ways of living in nature and Eurocentric sciences. But if we are aware of these potential pitfalls, comparisons can respectfully bridge Indigenous and scientific ways of knowing nature. Due to space limitations, the comparisons in this chapter have been simplified and appear in the context of Chapters 3 to 6.

The comparisons have immediate implications for school science. Science teachers can be a catalyst for fostering (1) open-mindedness, (2) dialogue that respects multiple perspectives, and (3) insights gained from dialogues and other experiences that encourage fine-tuning or reformulating what one understands and believes. Regardless of a student's cultural heritage, all students will be enriched when they learn something about other people's ways of knowing.

Comparisons

The following comparisons of IWLN and ES are offered with the deepest respect for each group. Our intention is to enhance a teacher's critical thinking when reading science teaching materials that discuss these two ways of knowing nature.

Similarities

Indigenous ways of living in nature and Eurocentric sciences have fundamental commonalities. They originate in the human impulse to make sense of the world so humans can take care of themselves. Moreover, both knowledge systems are based on observations gathered systematically. Thus, they share intellectual processes such as observing, questioning, interpreting, looking for patterns, inferring, classifying, predicting, verifying, problem solving, adapting, monitoring, and so on. Snively and Corsiglia (2001) describe several situations in which IWLN and ES have worked collaboratively: "Concerned with the multiple perils faced by their Nass River salmon, the Nisga'a [First Nation] have themselves implemented a salmon protection project that uses both the ancient technology and wisdom practices, as well as modern statistical methods of data analysis to provide more reliable fish counts than electronic tracking systems" (p. 19).

Elders and scientists are trusted practitioners in their own cultures. They exercise rational thought. In addition, scientists demonstrate communal characteristics of teamwork and consensus making, while Elders engage in collaborative discussion and consensus making. Both groups have in common such values as honesty, perseverance, inquisitiveness, and open-mindedness (Cajete, 2000b), as well as a passion for the aesthetic beauty of the universe.

Both groups employ their observations and empirical rationality in creative, intuitive, and precise ways. Tools are required to accomplish this. Indigenous technologies include those invented in the past and those appropriated from Eurocentric science and technology, along with "the preparation of the mind and heart" (Cajete, 1999, p. 85). Scientists' tools extend human perception and measurement, often making possible indirect data concerning theoretical entities.

Both groups generate cognitive models. For Indigenous people, the process of developing models requires more than strengthening one's intellect; it involves one's spiritual, emotional, and physical capacities. It is about becoming whole and complete, as opposed to an intellectual process of creating mental models based on inert facts, statistics, and numbers. For IWLN, models concern wisdom-in-action. "Models include symbols, numbers, geometric shapes, special objects, art forms, songs, stories, proverbs, metaphors, structures and the *always present circle*" (Cajete, 1999, p. 85, emphasis added). The Circle-of-Life model can take the form of the Medicine Wheel (Dyck, 1998). Models in ES are also culture-based metaphors, but they are invariably mechanistic, often mathematical, and exclusively related to the intellectual domain.

The outcomes generated by both groups are being continually revised in light of new observations and new ideas. They have evolved in response to changing circumstances and are continuing to evolve today. Thus, both Elders and scientists can change their understanding of Mother Earth and the physical world respectively; thus, their understandings are dynamic and tentative. The IWLN taught in schools will most often be contemporary Indigenous knowledge, and less often pre-contact knowledge.

The outcomes from both groups are most accurately communicated in the language of each culture. For IWLN, it usually is an oral Indigenous language versed in sophistication and precision. For ES, the language is usually written in technically sophisticated and precise text adhering to the vocabulary and syntax of a specific paradigm.

In summary, both IWLN and ES exhibit

- rational, intuitive, and logical thinking;
- systematic empirical approaches, including experimental-like procedures;
- communal social structures;
- intellectual processes such as observing, inferring, predicting, and so on;

- a dynamic evolution of their ways of knowing nature; and
- cultural bias and cultural relevance.

However, the most basic characteristic they share is that both are anchored in culture, even though many people believe Eurocentric sciences are culture-free.

Differences

Although both knowledge systems have evolved over time, they differ in the duration and nature of that evolution. IWLN have ensured Indigenous peoples' survival for tens of thousands of years, often under adverse conditions. ES, formerly known as natural philosophy, have inspired and escalated technological and economic growth over the past 400 years since Galileo's time.

Although both knowledge systems share intellectual processes, differences appear when we examine *how* each cultural group enacts these processes.

- *Observations*: Scientists are expected to remain detached and disconnected from the natural world in which they are observing, although many are aware that their presence may affect the outcome of their experiment. This detached view conflicts with Indigenous worldviews, in which people see themselves as an extension of observable things and events.
- *Predictions*: Indigenous people hold a common view that we can never know completely all there is to know about the natural world. Learning cultural teachings is a lifelong endeavour that will never understand the Great Mystery, a metaphor for everything there is to know. On the other hand, culture-based predictions in ES are usually about finding out which ideas best represent how the world works, and thereby have the potential to exercise power and control over nature.
- *Methodologies*: Indigenous people have diverse methodologies complete with ethics, protocols, ceremonies, and prayers. There is "a right way of doing things"—a phrase that is often used by Elders. Indigenous ways of passing on knowledge include storytelling, sharing circles, wilderness excursions, hands-on-activities, dances, songs, drama, arts, crafts, and ceremonies. All these reinforce a worldview of interconnectedness. Culture-based empirical methodologies in IWLN emphasize holistic, monist, and spiritual power, whereas the culture-based empirical methodologies of ES usually emphasize the power of reductionism and Cartesian dualism.

Differences between IWLN and ES are also apparent when scientists listen to the historical oral knowledge of Indigenous peoples. "Events occurring over several generations may be condensed into a single generation. This limits the possibility that scientists [who assume rectilinear time] can date [a] scientific phenomenon on the basis of native tradition [that assumes cyclical time]" (Snively & Corsiglia, 2001, p. 16).

Another fundamental difference between scientific and Indigenous ways of knowing nature is the fact that ES express a community-based *intellectual* tradition of thinking that produces knowledge, while IWLN express a community-based *wisdom* tradition of thinking, reflecting, living, and being. Broadly speaking, an intellectual tradition emphasizes individual, critical, and independent cognition as exemplified by Descartes' famous dictum "I think; therefore I am." A wisdom tradition emphasizes group-oriented ways of being, exemplified by "We are, therefore I am" and by living in harmony with Mother Earth. The poet T. S. Eliot appreciated the distinction when he asked, "Where is the wisdom we have lost in knowledge?" (1963, p. 161).

Wisdom or wisdom-in-action is central to Indigenous ways of living in nature. Elders are revered for their wisdom-in-action. A scientist can certainly behave in wise ways by living a good life like anyone else. However, when scientists collectively determine scientific truth by following ES's intellectual tradition and stringent set of scientific values, normative wisdom will be absent in most scientific paradigms. For example, the scientific truth of Newtonian laws of motion or the Krebs cycle in biology gives no indication of how nature should be treated with respect. Normative wisdom can certainly be a *value-added* feature to scientific truth, for example, when determining what policy should be followed in response to a scientific conclusion. On the other hand, holistic, relational, and spiritual Indigenous ways of living in nature *innately* guide community action toward treating nature with respect; for example, "Something remarkable is happening in Indian country: Tribes whose lands were once taken from them are setting an example for how to restore the environment" (Bowden, 2010, p. 81).

Most scholars agree that IWLN and ES are both rational, but their rationality is culture-laden and differ in several ways. These differences are summarized in Table 7.1. The categories in this table do not represent separate isolated ideas. Repetition occurs because some ideas belong in two or more of the categories. The table synthesizes cultural differences discussed in Chapters 3 to 6. These cultural differences either highlight strengths or acknowledge limitations, depending on one's point of view and on the specific situation in which the knowledge is used. For example, in the category "Association with human action," the characteristics of IWLN may have an advantage over ES in many resource management deliberations, but in the category "Type of validity," the characteristics of ES will certainly have an advantage over IWLN in determining energy flow in an industrial system. The table illustrates the complementary ways IWLN and ES deal with nature, but it does not express a simplistic dichotomy.

Table 7.1. Comparisons of Indigenous ways of living in nature and Eurocentric sciences*

General perspective	Monist, spiritual, relational, and intuitive descriptions/explanations of nature; found in the wisdom tradition of thinking, reflecting, living, and being *compared with* Dualist, materialist, comparatively non-relational, and often mechanistic descriptions/explanations of nature; found in the critical, independent, intellectual tradition of thinking
Social goals	Communal wisdom-in-action for the survival of the group, family, or community *compared with* An individual's scientific credibility, and many other social goals defined by the context of the scientific work, such as medical advances, environmental crises, and progress in a Western capitalist society
Assumptions	Mother Earth is mysterious and in continual flux *compared with* Nature is knowable and constant, but nature changes in consistently knowable patterns
Intellectual goals	Coexistence with the mysteries of Mother Earth by celebrating mystery through the maintenance of a host of interrelationships *compared with* Elucidation of mystery by describing and explaining nature in ways acceptable to a community of scientists

Fundamental value	Harmony with Mother Earth by balancing a web of interrelationships for survival *compared with* Power and dominion over nature by understanding how nature works
Association with human action	Intimately, subjectively, morally, and ethically related to human action with respect to the seven generations that came before, and the seven yet to come *compared with* Formally refrains from normative prescriptions of human action, which deal with subjectivity
Notion of time	Cyclical—there is no beginning and no ending *compared with* Rectilinear
Concepts of knowledge	Holistic, relational, and place-based *compared with* Reductionist, anthropocentric, and generalizable
Type of validity	Content validity as suggested by Aristotle's notion of intelligible essences, and supported by tens of thousands of years of survival based on that content *compared with* Predictive validity in anticipating observations accurately; the cornerstone of natural philosophy and Eurocentric sciences for about 400 years
Learning goals	Learning to become whole and complete (mentally, spiritually, emotionally, and physically) *compared with* Understanding a repository of knowledge developed in a linear reductionist manner with a primary focus on intellectual and physical pursuits
Socio-political context	Devalued/destroyed through colonization and oppressive globalization or by an ideological political revolution *compared with* Held up as an icon of prestige, power, privilege, and progress

*Notable exceptions to this summary exist, as discussed in Chapters 4 and 6.
Without the context of Chapters 3 to 6, this table might be misconstrued as a simplistic dichotomy. Comparisons are complex and nuanced.

Another point about the differences between ES and IWLN arises from history. From the sixteenth to the twentieth centuries, colonization by means of residential schools, racist policies, and religious intrusion, devalued and sometimes destroyed Indigenous ways of living in nature. Some national governments used ES as a tool to reinforce the oppression and colonization of people who embraced IWLN. The dominance of ES in the world today is in part directly related to past colonization and to globalization in the twentieth and twenty-first centuries. Globalization continues to encourage the devaluation and destruction of Indigenous knowledge worldwide (Davis, 2009; Knudtson & Suzuki, 1992; Sillitoe, 2007). But science teaching enhanced with an Indigenous perspective seeks to avoid these *neo-colonial* practices that continue today in subtle ways. These subtleties include inadvertent stereotyping such as referring to Indigenous people as "other," or the unintentional use of offensive names such as "Eskimo." These are very noticeable to most Indigenous people but not to many non-Indigenous people. Neo-colonial oppression is also illustrated by the marginalization of Indigenous students in school and university science programs. ES have been privileged over IWLN by the political, economic, and military dominance of colonizing and globalizing nations or by ideological revolutions such as a Marxist coup (Davis, 2009). This socio-political context illustrates a vast difference of power between ES and IWLN; therefore, it appears in Table 7.1.

A Scheme for Comparisons

Many cross-cultural science teaching materials present comparisons between Indigenous knowledge and Eurocentric science in a two-column table, which emphasizes their differences. This simple dichotomy is to be avoided, if possible. Table 7.2 avoids it by using a four-column format that highlights the *common ground* between IWLN and ES. Although this format unavoidably simplifies the diversities and complexities within both IWLN and ES, it still provides a useful overview. Consider the table as a work in progress.

Table 7.2 is organized around the following four main themes:

- general ideas about reality
- ideas about how to understand the world, and what those ideas are based on
- what is valued within each knowledge system
- issues of power associated with each knowledge system

These themes correspond roughly to what academic scholars refer to as the ontology, epistemology, axiology,[21] and political status of IWLN and ES. Within the second theme is a sub-theme, "Empirically Based," indicated by the shaded portion of Table 7.2. A degree of repetition in the table reflects that some ideas are applicable to more than one theme and category.

Some Elders speak about "two-eyed seeing," a perspective that emphasizes the strengths of both Indigenous ways of living in nature and Eurocentric sciences (CBU, 2007; Hatcher, Bartlett, Marshall, & Marshall, 2009a-b; Marshall, 2007). (For further discussion, see Chapter 8, the section "Deborah's Story Revisited.") Table 7.2 should be examined from the perspective of two-eyed seeing.

Classroom Applications

IWLN and ES coexist and share important common ground, which should be emphasized in school science. For example, both knowledge systems affirm that soil is alive. For Elders, soil is alive with Spirit; for scientists, soil is alive with insects and micro-organisms unless the soil has been sterilized.

When explaining the birth and death cycles of stars and their associated nuclear fusion, a teacher can easily identify a fascinating case of common ground. Eurocentric astronomers describe atoms being produced during the life cycle of stars. They say smaller atoms join together to make larger atoms, in a process of nuclear fusion. And, when most stars eventually die, they blow themselves into space dust. Our solar system is thought to have been formed out of such space dust, produced by a few generations of stars. From a scientific viewpoint, this means that humans and our Earth are made of matter (atoms) produced in stars that exploded long ago. In a material way, we literally come from stars. The same idea is also seen in the quote in Chapter 6 (the section "Relational") from Carol Schaefer (2006, p. 146): "… we are all cosmic beings. We come from the stars." These parallel ideas provide a concrete example of common ground between the two knowledge systems. The basic idea that humans come from the stars is shared, although the exact details differ.

Another fundamental Indigenous idea affirms that we are all related to everything in Mother Earth (see Chapter 6, the section "Relational"). This idea has richer meaning when we understand scientifically that Earth and

---

[21] Ontology refers to general ideas about what reality is. Epistemology refers to ideas about knowledge, how we know the world, and what those ideas are based on. Axiology generally refers to values, morals, and aesthetics; but, in this context, we focus only on values.

all the atoms in it, come from the same primordial stars in the Milky Way. We are all related by sharing the atoms produced by the same earlier stars.

Conclusion

As two-eyed seeing implies, people familiar with both knowledge systems can uniquely combine the two in various ways to meet a challenge or task at hand. In the context of environmental crises alone, a combination of both systems seems essential (see Chapter 2, the section "Enhancement of Human Resiliency").

In terms of educational learning theory, students can produce 'hybridized knowledge' from what they learn from both systems, creatively combining parts of each as a situation may require (Sillitoe, 2007; van Eijck & Roth, 2007). From a different theoretical perspective, students can achieve 'secured collateral learning' (Aikenhead & Jegede, 1999). Culturally diverse knowledge systems are resolved by idiosyncratically making sense of how they are related, according to a person's worldview. Hybridized understanding and secured collateral learning are theoretical frameworks for explaining some outcomes of cross-cultural learning. An alternative to hybridized understanding or secured collateral learning can be experienced by some people who feel more comfortable drawing on only one of the two knowledge systems at a time, depending on the situation. This is called 'parallel collateral learning' (Aikenhead & Jegede, 1999) and neither knowledge system is rejected.

The coexistence of Indigenous ways of living in nature and Eurocentric sciences can occur in school science and in out-of-school situations. When dealing locally *and* globally in a classroom with issues such as sustainability and ecosystems, the two knowledge systems together are more versatile and effective for students than one knowledge system alone (Clark & Dickson, 2003; Sillitoe, 2007; Snively & Corsiglia, 2001). For out-of-school situations, ES might help solve a local community problem, when appropriate, while local place-based knowledge can help ES achieve a much richer understanding of a particular geographical place and a much broader understanding of our planet.

Table 7.2. Comparing Indigenous Ways of Living in Nature and Eurocentric Sciences

Themes	Indigenous Ways of Living in Nature[1]	Common Ground	Eurocentric Sciences[2]
General ideas about reality		the impulse of humans to make sense of the world and to care for themselves	
	anchored in Indigenous cultures	anchored in culture	anchored mostly in Eurocentric cultures, but open to anyone immersed in the culture of Eurocentric sciences
	Mother Earth is mysterious and always in flux		nature is knowable and is both constant and changing in consistent ways
	unites physical and metaphysical worlds (monism)		mostly limited to the physical world of Cartesian dualism
	presupposes reality is spiritual		
	everything is subjectively interrelated, and these relationships create responsibilities		everything is assumed to be objectively related, but without associated responsibilities
	a pervasive belief that humans are equal to, or of less importance than, all other creations in Mother Earth		a pervasive belief that humans are the most important creation on the planet (anthropocentrism)

[1] These are largely indicative of Indigenous groups. They do not prescribe group characteristics, beliefs, or values.
[2] These are general descriptions. Exceptions within ES certainly exist, as indicated in Chapter 4.

Table 7.2. Comparing Indigenous Ways of Living in Nature and Eurocentric Sciences (continued)

Themes	Indigenous Ways of Living in Nature	Common Ground	Eurocentric Sciences
Ideas about how to understand the world, and what those ideas are based on		rational, metaphorical, dynamic, tentative over time	
	expresses a wisdom tradition of thinking, reflecting, living, and being		expresses an intellectual tradition of thinking
	emphasizes what nature **is**		emphasizes **how** nature works
	place-based		generalizable
	expresses a web of relationships in all creation		expresses an objectivity in being unrelated to what is being described or explained
	generally assumed to be the truth—a rich integration of multiple viewpoints		assumed to be the best representation of reality within a paradigm
		Empirically Based	
	emphasizes holistic, monist, spiritual power		usually emphasizes the power of reductionism and Cartesian duality
	accumulated observations over many generations		field studies or investigations into changes over great periods of time
	based on intuitive subjectivity		based on hypotheses and model building with the least subjectivity humanly possible

Themes	Indigenous Ways of Living in Nature	Common Ground	Eurocentric Sciences
Ideas about how to understand the world, and what those ideas are based on		*Empirically Based (continued)*	
	"experiments" rely on natural environmental changes over many generations and involve collective wisdom	experimental	experiments are undertaken by humans over relatively short periods of time
	source: both material and non-material worlds (monism)	observing	source: material world only (dualism)
	relationship based		no relationship exists between the observer and the object
	What is the world?	questioning	How does the world work?
	Who is that? How are they related? Who is doing that? What do they do?	classifying	What grouping does that fit? What are similarities/patterns that can arbitrarily define a group?
	to balance interrelationships in Mother Earth	predicting	to explain and predict natural events
	a myriad of representations of wisdom-in-action	models	mechanistic, often amenable to quantification, many are visual
	conducted by everyone, includes personal experience	monitoring	conducted by professionally trained experts and technicians
	appropriated tools and processes from modern technology; pre-contact tools and processes; "listening" to nature; preparing the mind and heart spiritually; and stories and art forms passed down in the oral tradition	technological tools and processes to investigate nature	advanced technologies and scientific processes approved in paradigms

Table 7.2. Comparing Indigenous Ways of Living in Nature and Eurocentric Sciences (continued)

Themes	Indigenous Ways of Living in Nature	Common Ground	Eurocentric Sciences
	Elders' collaborative discussions provide advice on what to do, based on experiences passed down and through one's own life.	communal	Scientists work in teams, invisible colleges, and paradigms to decide (via argumentation and consensus making) what to believe as "true."
	wisdom-in-action for the survival of plants, animals, and humans	social goals	an individual's scientific credibility; plus environmental issues, corporate profits, medical advances, national security, economic progress, and knowledge for its own sake, among other goals
	coexistence with mystery through maintaining a host of interrelationships	intellectual goals	elucidation of mystery by describing and explaining nature in ways acceptable to a community of scientists
	content validity suggested by Aristotle's notion of intelligible essences, and supported by tens of thousands of years of survival based on that content	validity	predictive validity indicated by anticipating observations accurately; the cornerstone of natural philosophy and Eurocentric sciences for the last 400 years
	local, oral Indigenous language, which is technically sophisticated, precise, and place-based	dissemination of ideas (when appropriate)	written text, which is technically sophisticated and precise, and which adheres to the vocabulary, syntax, and genre specific to a paradigm
	cyclical, with no beginning and no ending	time	rectilinear

Themes	Indigenous Ways of Living in Nature	Common Ground	Eurocentric Sciences
What is valued within each knowledge system	flux	honesty perseverance inquisitiveness open-mindedness logic curiosity aesthetic beauty creativity intuitiveness precision repeatability	consistency
	intimate, subjective, moral, and ethical; related to human action with respect to seven generations back and seven generations yet to come	human action	formal and "objective;" does not deal with normative prescriptions of human action
	relationships with, and responsibilities to, everything in creation; high subjectivity		"objective" disinterest between the observer and what is observed and interpreted, as far as is humanly possible; low subjectivity
	harmony with Mother Earth for survival; stewardship		power and dominion over nature for materialistic progress, political power, healthy well-being, and academic curiosities

Table 7.2. Comparing Indigenous Ways of Living in Nature and Eurocentric Sciences (continued)

Themes	Indigenous Ways of Living in Nature	Common Ground	Eurocentric Sciences
	sustainability		either neutral to sustainability or an entitlement to progress
	generosity		profit, progress, and credibility motives
	wisdom		understanding, knowledge and processes
	collaborative		competitive
Power imbalance	subordinated by dominant/oppressive/colonizing cultures		held in high esteem by dominant/oppressive/colonizing cultures
	a basis of self-identity, resilience, and resistance by Indigenous peoples		a basis of professional identity

CHAPTER 8

Building Bridges of Understanding: General Advice for Teachers

This chapter provides general ideas for science teachers who want to create a science classroom where students can learn the best of both Indigenous and scientific ways of knowing nature. These ideas are suggestions and do not provide a panacea for all situations. They serve as a starting point for thinking about culturally responsive school science, as well as how to incorporate Indigenous knowledge[22] into science classrooms. Although there is no universal way to combine Indigenous knowledge and Eurocentric sciences in school science, important cultural considerations can serve as a guide for combining the two (Michell, 2007). These considerations are explored in this chapter by discussing the following topics: resources for science teachers, Indigenous student learning, the classroom environment, instructional approaches, student assessment, the importance of Indigenous languages, and teachers' expectations of Indigenous students. The chapter also includes a further analysis of Deborah's story (Chapter 7), and ends with Mr. Chang's reflections on a challenging question posed by an Indigenous student in his science class.

Teachers are complex human beings who are gifted in their own unique ways. Therefore, a teacher's inner wisdom is an excellent resource to draw upon. However, it is also imperative for a teacher to be aware of the broad socio-political and historical issues affecting the lives of Indigenous people,

---

[22] Given the fact that the meaning of Indigenous ways of living in nature (IWLN) has been firmly established in this book, Chapter 8 follows the convention of using "Indigenous knowledge" as a synonym for IWLN.

and to be mindful of the history of Indigenous education in countries where colonization has occurred. These issues include the effects of this colonization on communities and schools, particularly on their culture, ways of knowing, language, values, and traditions in postcolonial school science. The term "postcolonial," as defined in Chapter 6 (the section "Conclusion"), means that efforts are being made to diminish and extinguish the power of colonialism. A postcolonial attitude helps teachers and administrators avoid misunderstandings when including Indigenous knowledge in school science (Barnhardt, 2006).

To ignore a postcolonial point of view is to risk taking on a neo-colonial role by subtly and even unconsciously devaluing Indigenous knowledge (see Chapter 7, the section "Differences"). For instance, some scientists work on commercial projects in a social context of corporations or governmental agencies, which are often related to some agribusiness or resource industries. Many of these projects result in the physical devastation of Indigenous lands, as well as a people's social and spiritual dislocation. Not so subtle neo-colonialism has been documented, for example, by Davis (2009), Settee (2000), and Sillitoe (2007). Individual scientists may not personally agree with their employer's neo-colonial role, but given the resulting oppressive consequences for some Indigenous people, scientists who participate in such projects must be associated with this neo-colonial behaviour. Davis (2009) drew attention to a contemporary case that remains unresolved in the courts at this time. At the headwaters of three major salmon-bearing rivers in British Columbia (Stikine, Skeena, and Nass), and "against the wishes of all First Nations, the government of British Columbia has opened the Sacred Headwaters to industrial development" (p. 117). These First Nations include the Gitxsan and Nisga'a Nations, while the industrial development involves multinational resource corporations. The conflict arises in part from a non-Indigenous concept of land. "We accept it as normal that people who have never been on the land, who have no history or connection to the country, may legally secure the right to come in and by the very nature of their enterprises leave in their wake a cultural and physical landscape utterly transformed and desecrated" (pp. 118–119). These remarks are very much a reflection of an Indigenous consciousness. Through an awareness of such modern-day experiences of Indigenous peoples, a science teacher will be better prepared when an issue arises in class concerning an Indigenous nation in conflict with corporate enterprises and government agencies. Building bridges of understanding includes heightening a teacher's awareness of current instances of neo-colonialism.

Resources for Science Teachers

A variety of resources are identified here, along with advice about adapting them to a teacher's repertoire of practical knowledge for teaching (Michell, 2007).

Elder Involvement

Elders are essential teachers of community-based Indigenous knowledge and wisdom. They may serve as authoritative sources in the development of an enhanced science curriculum and for cross-cultural science lessons and units. Elders and other Knowledge Keepers[23], such as trappers, hunters, artisans, cooks, and traditional land users, are typically the carriers of language, values, and worldviews, as well as ways of knowing, living, and being. To gain access to their knowledge, a science teacher could first develop a relationship with a key person in a community. That person could introduce the teacher to other people who possess the local, place-based, cultural knowledge the teacher needs to learn. This is what Ms. Smith, the Grade 8 teacher, did with an Indigenous bank teller (see Chapter 1, the section "A Vignette"). (As discussed below, some schools may have a designated Elder.) Teachers might 'adopt' a key person as a *Kookum* or *Mooshum*, the Woodlands Cree words for Grandmother or Grandfather, someone who has a collective memory of the community.

Follow the local protocol of offering a gift when requesting such knowledge from an Elder (Saskatchewan Indian Cultural Centre, 2009). The acceptance of a gift signifies a beginning of a relationship. Appropriate gifts include tobacco, tea, home-made jam, some type of service, a blanket, or any other accepted way of honouring an Elder's teachings. If an Elder comes to a school, arrange transportation for him or her, and, if possible, provide an honorarium. If in doubt about local protocols in a community, ask the Elder what the protocol is. They will be happy to tell you because you are seeking to be informed, which expresses respect for their culture. In subsequent requests for help or information, offering a gift expresses your respect for the Elder and for your relationship with him or her.

In some school communities such as those in large cities, Elders may not be available. Their numbers are small, their responsibilities are great. Alternatively, other culture experts, who call themselves by various names

---

[23] The title "Knowledge Keeper" is defined in Chapter 5 (the section "Clarification of *Knowledge* and *Nature*").

such as Knowledge Keepers, Traditional Knowledge Advisors, Traditional Knowledge Trustees, and so on, can be very helpful.

If you do not have contacts like Ms. Smith had, how do you know who is an Elder? Some schools or school districts have a list of designated Elders willing to help teachers. Some school districts have an "Aboriginal Consultant" acting as a liaison to a local Indigenous community. Tribal Councils and other Indigenous organizations may help identify an Elder as well. Do not be surprised if, on first contacting an Elder, he or she humbly says that they do not know very much. They are teaching the value of humility—we humans know so very little when there is always so much more to learn. Elders do not like to be placed on a pedestal or have people think they are an expert or superior to others. "We are all equal," they often say.

Several visits with an Elder may be necessary to get to know them before asking questions about Indigenous knowledge and wisdom. Patience is important as relationships take time to build and an Elder's transmission of knowledge does not follow contemporary Eurocentric timelines. An Elder will *intuitively* know when the time is right to talk to you about the knowledge you seek.

Collaborate with individual Elders or Knowledge Keepers to identify key concepts and content relevant to the particular community's context. For example, a teacher might show an Elder or Knowledge Keeper a page from the science curriculum or from a textbook that conveys Indigenous knowledge identified as belonging to another community or nation. Then ask, "If this is what that Indigenous community generally believes, what should *we* also teach here in our school?"

Elders and Knowledge Keepers can be invited into a classroom to enhance students' learning experience, provided the person feels comfortable being in a school. Residential school experiences can have lasting negative effects on how people feel about schools, generation after generation. If a classroom visit is not possible, these resource people might meet with students outside of school at a convenient place such as a park or community event. Resource people may also be contacted by some students during out-of-school hours for information and ideas. If so, these students must be taught the local protocols of gaining knowledge and interviewing people. Students must ask an Elder what knowledge can be repeated orally, and what can be recorded for others to read. Occasionally, an Indigenous student or two might be matched with a Knowledge Keeper in an outreach program. For example, the student or students might visit a trapping line with a Knowledge Keeper, creating a context for comfort zones to be attained, as well as imparting a valuable experience.

Whether in classrooms or out-of-school settings, prepare students to show the utmost respect for an Elder. Reinforce the Indigenous knowledge that students learn, and teach them how to document this knowledge, if appropriate, using digital technology along with short written explanations or anecdotes. Students could then report what they learned to the class, if permission has been given by the Elder or Knowledge Keeper. Additionally, computer classes can teach website construction skills so students can share their public knowledge with others around the world.

When Elders speak in classrooms, they talk holistically. They will express wise and proper ways of living, often related through storytelling. For some science teachers, this genre of presentation might at first be seen as wandering away from the specific topic identified by the teacher. However, knowledge *and* wisdom exist holistically in Indigenous ways of thinking, unlike that of Anglo- or Franco-Eurocentric cultures. The teacher's initial assessment of the presentation may result in a cultural misunderstanding. But the teacher's patience, respect, and reflection can often resolve such misunderstandings.

Elders and certain community people in general are gifted in different areas. Their skills might include those of a plant gatherer, herbalist, storyteller, leather maker, beader, ceremony keeper, birch bark canoe maker, linguist, trapper, hunter, or midwife. Most will be pleased to offer ideas and to review or to endorse a teacher's activity-centred lessons that connect curriculum/textbook content with the community's culture. Students can then use the Internet to extend their learning about other cultural groups in order to break down barriers and to enrich their understanding. The goal is to learn about other ways of knowing while developing relationships with "all our relations" in Mother Earth.

Teachers must be aware of cultural customs, beliefs, and taboos that may conflict with some students' learning school science, such as the animal dissections described in "Deborah's Story" in Chapter 7. Elders or Knowledge Keepers can be instrumental in resolving such conflicts. They might teach students about protocols, ethics, rituals, ceremonies, traditional values, and how these are connected to proper ways of hunting, collecting, dissecting, handling, and viewing animals and plants. Cultural teachings and protocols have the potential to enhance learning regardless of a student's cultural background. This is because Indigenous teachings place heavy emphasis on both stewardship and sustainability, important concepts in science curricula for all students.

Plan 'medicine walks' with an Elder or Knowledge Keeper to connect students with the land. For example, students can learn about plants used for food, healing, and ceremonies. Before such excursions, students should learn the protocols for picking plants and learn how these protocols are connected to Indigenous worldviews.

A teacher learns along with his/her students, and so the teacher's role changes from an instructor to an adult learner, thereby providing a living example of lifelong learning for students.

Community Contexts

Taking part in cultural camps, community gatherings, seminars, community hunts, and ceremonies such as powwows, helps teachers who are now adult learners experience the collective worldview of an Indigenous community (Saskatchewan Indian Cultural Centre, 2009). At a community cultural event, a teacher should only share knowledge when invited to do so. An Indigenous worldview can provide a framework for planning a science lesson containing Indigenous knowledge. For example, when planning science lessons, teachers will become aware of the power of the number four and can use that awareness when considering the emotional, physical, mental, and spiritual development of their students.

As with Elders and Knowledge Keepers, the local community can be a rich resource for teaching school science in many ways. When feasible, involve parents, Indigenous language instructors, and other community members in planning, developing, implementing, and evaluating science instruction based on their local knowledge. This culturally responsive action focuses classroom attention on students' cultural backgrounds very effectively, helps augment students' resiliency, and ultimately causes significant improvement in their educational experience and success (Reyhner, 2006; Sutherland, 2005). To make school more relevant, introduce local artifacts along with community events that connect with the science curriculum. For example, develop lessons that revolve around seasonal cycles and around particular Indigenous technologies, both contemporary and traditional. Introduce and explore a scientific perspective on, for instance, snowshoes, canoes, hunting tools, fishing implements, or Indigenous musical instruments, as well as the production of food, clothing, art, and utensils.

Always instill a strong sense of community, akin to an Indigenous culture's collective orientation expressed as "We are, therefore I am." Instill the value of giving back to one's community. This can be accomplished by students engaging in science-based activities that directly benefit their community, for example, designing a recycling system that consciously considers seven generations yet to be born.

Have Indigenous students investigate their own communities by undertaking small-scale research. For example, they can find out how Indigenous values are relevant in the practice of Eurocentric sciences. Introduce problem-solving circles, comprised of a group seated around a circle usually with a space to break the circle. Problem-solving circles can address scenarios that reflect community realities. A conscious connection to students' communities will deepen their understanding of themselves, as well as the ways that Eurocentric sciences intersect with their communities.

Teach Indigenous and non-Indigenous students the proper way to interview community people, to discuss the results of these interviews, and to relate their findings to other information. If applicable, prepare students how to properly ask permission to record these events in either audio or video format. Have students share their experiences through varied modes of communication. Show them how to "publish" their research results and how they might disseminate what they learned. Perhaps organize a 'knowledge festival' or 'knowledge fair' at which students present their research to the community (see details below in the section "Other Approaches to Teaching"). If Indigenous students are reluctant to speak in public, they can develop their speaking confidence by participating in small projects throughout the year.

With people who are fluent in the local Indigenous language, plan Indigenous 'science linguist' camps. Focus on learning constructs and comparisons from the viewpoint of an Indigenous language, and on learning how these constructs and comparisons reveal different understandings related to the natural world (Michell, 2007). These immersion camps might involve canoe and kayak trips (Kawagley, 1995). In the winter, draw on sled dogs, snowshoes, and toboggans, if available, to get an authentic sense of being out in the land as the ancestors were long ago. Use weekend gatherings, short excursions, one-day events, pond studies, swamp studies, fish studies, habitat explorations, and specific ecological observations. Have Indigenous students document their experiences with the goal of sharing what they learned with their community.

We offer here a few words of caution: In remote rural communities especially, mutual trust and a positive rapport must be established before a teacher prepares students to share Indigenous knowledge from their family or community. Establishing such a rapport begins by obtaining support from a key community member and by informing families what you want to accomplish, as Ms. Smith, the Grade 8 teacher, did (see Chapter 1, the section "A Vignette"). Elicit community suggestions about what to accomplish and how to do so. Each community is different. To locate more specific ideas on how to engage the local community, check with the school to see if it already has a cultural events plan or policies in place encouraging the involvement of community members. In addition, consult the document "Stories from the Field" in the *Rekindling Traditions* project (Aikenhead, 2000, online). Be sensitive to families strongly steeped in their own Indigenous tradition who may have taught their children to respect the fact that children know very little at this stage in their young lives. In such a family, it may not be a child's role to instruct or inform classmates and a teacher, for fear of misleading others who are also on their learning path. Thus, some students may be reluctant to share knowledge because they do not want to show disrespect toward teachers whose role is to teach. Furthermore, they may not want peers to see them as being knowledgeable, or even worse, as being different from their peers.

In multicultural urban settings particularly, be prepared for some Indigenous students to be self-conscious about sharing any Indigenous knowledge they may have, for fear of being singled out and questioned about their personal beliefs or cultural worldviews. Like most students, they too may prefer to fit into the "norm" set by their peers.

Role Models

Role models and mentors in science education are valuable assets, especially Indigenous role models (Environics Institute, 2010; Michell, 2007). Contact science-related professionals through renewable resource agencies and other sources to interact and build relationships with your science-proficient students. Demystify the work of scientists and engineers. Science-proficient students can investigate science-based occupations by communicating with employees in health, agriculture, renewable resources, and industry sectors.

Invite role models into your classroom. When speaking with these resource people ahead of time, be specific regarding what they are expected to talk about, so that the presentation connects with the lesson's objectives.

Have students write questions prior to the visit to ensure an exchange of relevant ideas. While it is important to invite role models who share your students' heritage, it is equally crucial for students to build relationships across cultures and learn respect for the perspectives of others.

Lastly, have students explore, describe, and share positive contributions made by Indigenous people in fields related to Eurocentric science and technology (*Native Access to Engineering Program*, Appendix C; Selin, 1992).

Teaching Materials and Resources

Be critically selective with teaching materials, books, websites, and other resources brought into the classroom. Resources transmit hidden values, norms, traditions, and stories of the cultures and subcultures of those who produce them. As described in Chapter 6 (the section "Mysterious"), for instance, some resources may be inappropriate by being insensitively Eurocentric. Be aware also of stereotypes, inaccuracies, or narrow perspectives depicted in teaching materials, but turn these shortcomings into teachable moments, if a resource is to be used at all. Transform a negative example into a positive learning experience for students. If in doubt about a resource, have it previewed by a reputable person from the Indigenous community, such as a cultural or language instructor. Some materials may need to be modified and complemented with resources that better reflect the local Indigenous community's perspectives.

Today, some science textbooks are becoming more inclusive of Indigenous knowledge. For some textbooks, Elders were engaged to choose the Indigenous content, and appropriate Indigenous processes and protocols were followed (Aikenhead & Elliott, 2010). Such texts include *Pearson Saskatchewan Science 6* (Johanson, Mohr, Treptau, Wallace, & View, 2009), *Pearson Saskatchewan Science 7* (Brockman, Doepker, Stephenson, Wallace, & View, 2009), *Pearson Saskatchewan Science 8* (Boulton, Brockman, Johanson, Wallace, & View, 2010), and *Pearson Saskatchewan Science 9* (Hounjet et al., 2011). In Canada's Northwest Territories, an online teachers' resource manual supports the Department of Education's Grades 10 to 12 cross-cultural science curriculum, *Experiential Science*, with units and student activities (NWT Protected Areas Strategy, 2009). See CRYSTAL (2010) for Nunavut.

Select books that accurately portray Indigenous people, preferably written by Indigenous authors. Indigenous students relate much better to books that describe events, people, and places with which they are acquainted. Teachers need to be cautious about using stories with subject matter unsuitable for children. Resource materials should always be screened. Teachers should also be mindful of local taboos. For instance, some Indigenous communities do not say "grizzly bear" out loud in their local language when walking a trail, to avoid inviting one into the camp. Elders and language specialists could be very helpful in screening materials and identifying taboos. The key purpose for interacting with stories is to explore and discuss the hidden messages and core teachings they contain. Indigenous storytellers will often choose a story to retell to suit a particular audience. An Elder's story is meant to cause people to reflect on the tale and to reach their own meaning, even if that takes months or years to achieve.

For North American teachers, we recommend Garvin (2005; *Carving Faces, Carving Lives*), Ipellie (2007; *The Inuit Thought of It*), Landon (2008; *A Native American Thought of It*), and Maryboy and Begay (2005; *Sharing the Skies: A Cross-Cultural View*). Tribal Councils and other Indigenous authorities sometimes publish relevant books, often entitled "Voices of the Elders" and "Stories from the People." In addition, students are always interested in the personal stories surrounding scientific events, but less so in any dry, impersonal accounts of the same events. Culturally responsive teachers tend to transform lectures into stories whenever possible.

To learn about best practices, lessons, models, processes, and general ideas, we suggest that teachers read professional literature about Indigenous science education. The ample reference section provided in this book offers a library of possibilities. Substantial practical day-to-day advice is offered by Pearson Education Canada (2010) in *Pearson Saskatchewan Science 6–8 Program Overview*. A special issue of the professional journal *Green Teacher* (Hatcher et al., 2009a) is dedicated to implementing cross-cultural school science for a variety of topics. Professional literature can help one understand, develop, and modify lessons and unit plans to reflect an Indigenous community's place-based knowledge.

Appendices A, B, C, and D in this book offer other specific resources:

- Appendix A summarizes a cross-cultural teaching unit, which illustrates bridging Indigenous and scientific ways of knowing nature.
- Appendix B guides critical reflection on this book's content by posing questions to consider and discuss.
- Appendix C offers a selection of useful websites.
- Appendix D is an annotated bibliography of primarily non-academic books that present an Indigenous perspective in a more detailed, personal, and, in some cases, humorous way.

Indigenous Student Learning

Coming to know is one way to describe learning within Indigenous cultures, an idea introduced in Chapter 5 (the section "Clarification of *Coming to Know*"). It signifies a personal, participatory, constructive process toward gaining community-based knowledge and wisdom-in-action. It requires that "each person learns for himself or herself through the processes of growing up in contact with nature and society; by observing, watching, listening, and dreaming" (Peat, 1994, p. 59).

Coming to know is a journey into deep understanding of events or of processes in daily life (Cajete, 2000b). For instance, if someone is coming to know a plant, they enter into a personal relationship with the plant in such a way that makes a person

> sensitive to the fact that each has its own energy. "Coming to know," or understanding the essence of a plant, derives from intuition, feeling, and relationship, and evolves over extensive experience and participation with green nature.... Native use of plants for food, medicine, clothing, shelter, art, and transportation, and as "spiritual partners," was predicated upon establishing both a personal and communal covenant with plants in general and with certain plants in particular. (p. 110)

Another way to describe learning within Indigenous cultures draws on the notion of 'learning styles.' Contrary to popular belief, however, there is little evidence that a stereotype learning style for Indigenous students exists. Instead, evidence points to *recurrent learning strengths* found among Indigenous students (Hughes, More, & Williams, 2004; Alaska Native Science Commission, 2009). These strengths include

- holistic more than analytic;
- visual more than verbal;
- oral more than written;
- practical more than theoretical;
- reflective more than trial-and-error;
- contextual more than decontextual;
- personally relational more than an impersonal acquisition of isolated facts and algorithms;
- experiential more than passive;
- oriented to storytelling sessions more than didactic sessions; and
- taking time to reflect more than quickly coming to an answer.

Although these recurrent learning strengths obviously embody cultural features of Indigenous ways of living in nature, they are present also in non-Indigenous students to varying degrees. The opposite is also true: some Indigenous students favour the recurrent learning strengths of science-oriented non-Indigenous students.

The learning strengths identified above are not generally rewarded in conventional science classrooms. School science usually emphasizes analytic, verbal, written, theoretical, trial-and-error, decontextual, impersonal, passive, and didactic learning to get the right answer quickly. Such recurrent learning strengths of conventional school science may frustrate or alienate students who do not share them. Thus, a variety of pedagogical approaches or differentiated instruction will resonate with a greater number of students in a science class. A classroom that encompasses the recurrent learning strengths of Indigenous students will likely help those students feel comfortable in learning school science and in becoming aware of how scientists think.

Yet, all students, Indigenous and non-Indigenous, should be encouraged to expand their repertoire of learning strengths to enrich their capacity to learn in more diverse ways. Rather than accepting a student's learning strength as the *only* way that student can learn, teachers can help holistic learners build complementary analytic strengths during particular science activities without expecting them to diminish their holistic learning strength. Similarly, non-Indigenous, science-oriented students with analytic, verbal, written, and abstract learning strengths will benefit from engaging in holistic, image-bound, and practical thinking in some specific circumstances in school science.

The Canadian Council on Learning (2007a) undertook a significant study of Indigenous learning that resulted in three "holistic lifelong learning models," one each for First Nations, Inuit, and Métis peoples. Each model uses stylized graphics to express relationships among three major features: the purposes of learning, the outcomes of learning, and the group's collective worldview. Interactive online models offer rich detail. The model relevant to a teacher's community would make a splendid colour poster for a science classroom—a great conversation piece for parents and visitors.

Lastly, never forget what some older Indigenous students say about what is most important to their learning: "You don't take a class; you take a teacher" (Hampton & Roy, 2002, p. 17). The teacher and his or her classroom are integral to what students learn.

Classroom Environment

Teaching a community's Indigenous knowledge in a science classroom requires teachers to foster a life-long learning approach by becoming students of that culture themselves. When teachers assume the role of student, they demonstrate a passion for learning and they model skills such as carefully observing, actively asking questions, and immersing themselves in local cultural activities.

When a teacher learns Indigenous knowledge from students who bring it from home, the classroom environment changes positively and pervasively, creating a new shared responsibility for learning. Overall, this builds reciprocal respect and increases students' willingness to accept the teacher's authority.

As much as possible, promote Indigenous students' ownership of their learning by allowing them reasonable input into what is learned, when it is learned, how it is learned, and how learning is assessed, which is also good advice for non-Indigenous students. Students develop leadership skills when given an opportunity to make meaningful decisions about classroom activities. These and similar strategies also work well for all students in science classrooms that are multicultural, urban, and elementary (Patchen & Cox-Petersen, 2008). Allow students to help plan field trips and excursions. For example, if they are in charge of planning and purchasing food, they quickly will learn the life lesson of natural consequences for decisions made and actions taken (see the teaching unit "Survival in Our Land" in *Rekindling Traditions;* Aikenhead, 2000).

Métis scholar Madeleine MacIvor (1995) encourages science teachers to build a sense of place and belonging in their classrooms, one steeped in humility, respect, and reverence for all life. The ideal classroom environment reflects the cultural diversity within each community. The idea is to foster a home-away-from-home learning environment by reinforcing the importance of relationships, kinship, and collective well-being (Hatcher et al., 2009b). Decorate classroom walls in a visually and culturally stimulating way with items such as dream catchers, bark baskets, and local herbs, as well as posters of Indigenous scientists, engineers, leaders, and Elders. Community resources can help here. Use Indigenous symbols, themes, colours, artwork, language concepts, and circular diagrams that show the interconnectedness of humans with the natural world. Have students create a circular collage of common animals and plants that surround their communities—bring the out-of-doors indoors. Hatcher and her colleagues

(2009a–b) offer further specific details on how to achieve a culturally responsive classroom environment.

It is also important to establish a home-away-from-home environment in urban multicultural classrooms (Barnhardt, 2006):

> While it is not possible to fully attend to the particular cultural needs of every student [in a culturally mixed school] on a daily basis ... the interests and strengths of each student can be recognized and rewarded through practices such as peer tutoring, cultural demonstrations, group projects, language comparison, etc. (p. 7)

The important point here is a teacher's recognition of and respect for students' cultural identities.

In all settings, there must be zero tolerance for racism, whether it is individual, systemic, or institutional racism. Indigenous students' experiences with racism contribute greatly to poor educational outcomes, especially in urban schools (Taylor, Friedel, & Edge, 2009).

Give student learning and well-being greater priority than test scores, when feasible. For any underachieving student, remember to distinguish between the good human being who inhabits that student's body and the marks that the same student earns. Improvement in the latter often occurs over time when teachers remind a student of the former. Respect and acceptance may not pay off immediately in improving a student's achievement during the school year, but a teacher's respect and acceptance can nonetheless underpin the student's later success in life (Wagamese, 2008).

Especially in higher grades, aim for a classroom environment that conveys the possibility of future occupations for students learning science. Provide opportunities for students to research careers. Have them create and display posters of Indigenous scientists, engineers, and technicians. Information resources on careers and exemplary individuals are found in Appendix C.

Become familiar with the local community's use of decision-making processes such as consensus and talking circles. Use these in the classroom as much as possible so students are able to express themselves in constructive ways while reinforcing the collective orientation of their community. Talking circles foster the expression of multiple perspectives, a critically important aspect in scientific problem solving. They can reveal students' prior knowledge and competencies that would have otherwise gone unnoticed using other teaching methods.

Also become familiar with the social conventions of how people in the community interact, including body language. For example, silence has many interpretations. Students' silence in a teacher's classroom may be frustrating for the teacher, but it may well be the way students show respect. Some students may be reluctant to answer a question if they feel that doing so may embarrass others who could not answer the question. Similarly, Elders seldom offer a negative response to a question. Their silence, therefore, can mean they do not want to say something negative about whatever they were asked.

Two issues may arise that will certainly affect a teacher's classroom environment. As discussed in Chapter 6, spirituality is a way of life for many Indigenous peoples; it cannot be separated from school and learning. "[Indigenous] spirituality is the sense of relationship or connection with all parts of Creation" (Coalition for the Advancement of Aboriginal Studies, 2002, p. 283). Religion, on the other hand, has been characterized by some people as a formally institutionalized form of worship. In some schools in Canada, religious instruction occurs, but in many schools it does not. Religious people certainly express spirituality; hence, religion and spirituality can be compatible, but they are not synonymous. Teachers, of course, should show respect for both the spirituality and religious affiliation of all students. Importantly, by making students aware of Indigenous spirituality, a teacher is not bringing religion into the classroom.

A second and related issue concerns the difference between understanding an idea and believing an idea, a distinction described in detail in Chapter 7 (the section "Deborah's Story"). An Indigenous student can understand Darwin's theory of natural selection without dismissing his or her belief in an Indigenous creation story. Similarly, a non-Indigenous student can understand Indigenous spirituality without believing it or dismissing his or her own religious beliefs. The following quotation illustrates how this is possible. It is taken from a Grade 8 science textbook: "According to the worldview of the First Nations and Métis peoples everything on Earth has Spirit flowing through it" (Boulton et al., 2010, p. 7). A science teacher should feel comfortable teaching the existence of Indigenous spirituality in order that students can *understand* it. Understanding, but not necessarily believing, is the teaching objective (Cobern, 1996). Such a teaching approach often resolves any fear some parents may have about Indigenous ideas, and it usually makes Indigenous students feel more included in the classroom. Teaching students *to believe* Indigenous spirituality is the role of families, communities, and Elders.

Instructional Approaches

Although Chapter 8 primarily focuses on Indigenous students, much of its advice about instruction and assessment applies equally well to non-Indigenous students. The advice will be familiar to successful teachers who give students relevant experiences that focus on meaningful learning rather than on simply memorizing vocabulary, explanations, and how to solve problems for tests (Aikenhead, 2006; Hutchison & Hammer, 2010). Meaningful learning engages students in understanding concepts well enough to make sense of phenomena.

Honour the knowledge, experiences, and skills students bring to the science classroom (Michell, 2007). Many Indigenous students are gifted with special cultural knowledge, abilities, and skills. Tap into these gifts by using them in a variety of instructional approaches and methods, also known as differentiated instruction.

The following general advice has guided experienced cross-cultural science teachers:

- Introduce a topic by drawing attention to something familiar to students.
- Build on students' existing knowledge by incorporating newer ideas at a pace comfortable to students.
- Provide students with ways to try out their new knowledge in different circumstances.
- Show that meaningful learning can be achieved when we learn from our mistakes.
- Plan time for students to work in-depth with their new knowledge and for them to learn from their mistakes.

Encourage students to feel comfortable about taking risks, such as talking in the "foreign" language of Eurocentric science (Kawasaki, 2002; Taconis & Kessels, 2009). Make them feel comfortable about being seen as a person who has learned some Eurocentric science without setting aside their Indigenous identity and community values. Role-playing techniques can be useful in these circumstances. Introduce laughter and humour into the classroom. This can be done by introducing games, competitions, demonstrations, drama, music, and art to make learning stimulating and motivating. Spontaneity and an element of mystery work wonders in science teaching. Surprise students during a class, just as the trickster or transformer would in Indigenous people's traditional stories. Keep your students wondering what you will do next!

Time-Honoured Indigenous Ideas about Teaching

Historically, Indigenous societies have rich and diverse ways for one generation to pass along knowledge and wisdom to younger generations. Key approaches include the following: demonstrations; supervised and unsupervised events and practices; sharing circles and storytelling; drawing, painting, singing, and making models; Elder guidance; apprenticeship; ritual and ceremony; dreaming and imagination; and intergenerational teaching (MacIvor, 1995). Indigenous people have their own worldviews and belief systems that underpin the way they teach their children. These traditional methods typically follow a pattern:

- Demonstrate what is to be learned while a child watches and listens intently. Repeat as necessary.
- Involve a child in small tasks if possible.
- Allow time for a child to reflect on and practice what is to be learned.
- Provide an opportunity for a child to show what they have learned, but only when they feel ready.

This pattern harmonizes with the recurrent learning strengths of many Indigenous students. Whenever feasible, learn more about this pattern by watching and talking about how people in the community teach their children. This knowledge may benefit classroom instruction from time to time.

Other Approaches to Teaching

Try a variety of teaching methods, incentives, materials, and contexts. What works in one context with one group of students will not necessarily work in another.

Rowland and Adkins (2003) summarized successful approaches to teaching school science to Indigenous students (mostly based on research in the United States):

- Integrate school science with other subjects (e.g., social studies, math, native studies) in line with holistic ways of learning. (See the extensive resource, *Integrating Aboriginal Perspectives into Curricula*, by Manitoba Education and Youth, 2003.)
- Make connections between the world of Eurocentric science and the world of the learner, so that school science is more practical for students.
- Explore differences between Indigenous knowledge and Eurocentric science, allowing students to reflect on these two ways of knowing

nature. In northern Manitoba, for instance, this worked successfully for Grade 7 Indigenous students (Sutherland & Dennick, 2002).
- Connect school science with topics and themes important to the Indigenous community, thereby giving a meaningful context to students' learning.
- Integrate a land-based ethic into instruction to underscore a student's web of relationships and responsibilities to the land.
- Use cooperative learning groups, which are more personal than didactic teaching, and which emphasize an oral tradition of learning.
- Instruct through experiential learning activities, whenever possible, to support students' *coming to know* school science content.

In addition, we add the following approaches:
- Begin any class with an attention grabber such as an anecdote or a question that draws on students' experiences related to the topic at hand.
- Ensure that students see the whole picture of what the lesson is about and the reasons why they are learning its content.

These general guidelines lead to more specific advice for all students. When students, usually in higher grades, have research papers or reports to complete, ask to see outlines and rough drafts early in the semester with the purpose of providing direction and tips. Get students to develop the body of the paper first before writing the introduction and conclusion. This helps them concentrate on the whole first, followed by the parts.

Incorporate a variety of visual aids when explaining concepts, processes, and information. Examples include rock samples, concept diagrams, charts, posters, slides, overheads, films, videos, models, photographs, cartoons, plants, and, where feasible, animals. Such aids harmonize with a visual recurrent learning strength. When out in the land, Indigenous children are taught to observe in multi-sensory ways and to cultivate their memory in order to survive. This multi-sensory approach should be encouraged in science teaching—the more input channels that are accessed, the more likely information will be retained. Different sensory channels provide alternative memory anchors from which students can access information.

In the classroom, make available a variety of scientific literature, newspaper clippings, models, charts, photos, tools, equipment, and other materials. Ensure they are visible or easily accessible to students. These props can make explanations much clearer for students whose recurrent learning strength is practical more than theoretical. Visual aids and other science teaching materials require adequate funds, which should be a high priority in the school's budget.

Provide opportunities in which students explore the historical development of Eurocentric science, its diverse methods, and its ways of thinking. Consider both the positive and the negative ways that Indigenous lands and resources are affected by projects in which Eurocentric science and technology have played a part. According to Nakawē (Saulteaux) Elder Danny Musqua (Johanson et al., 2009), positive examples include the helpful ways electricity from hydroelectric dams improves one's life. Negative examples include the adverse social and environmental impact of hydroelectric dams, mining, and logging. As a result of these damaging developments, some Indigenous families and communities have experienced intolerable changes in their way of life and living standards. Give students a chance to identify how Eurocentric science impacts their everyday realities today, and then expand this thinking to investigate how other Indigenous peoples around the world are affected by the same or similar impacts.

To learn about the people and the "real stories" behind scientific discoveries, have students read and discuss original journals, case studies, or narratives of scientists. Finding stories beyond the media stereotypes helps students to relate better to Eurocentric science and scientists. Provide opportunities to experience Indigenous methodologies for answering questions about the natural world. Students can then study and discuss the differences between the Eurocentric and Indigenous approaches, including how the knowledge was put to use in each cultural context.

Talk with students about values that guide a wide variety of scientific practices, determined by the diversity of social contexts in which scientific research is conducted (see Chapter 4, the sections "A More Realistic View of Eurocentric Sciences" and "Eurocentric Sciences Are Embedded in Social Contexts"). Compare and contrast these values that guide scientists with values that guide Indigenous community action. Find common ground and recognize differences. Students should be encouraged *to understand* the values that guide scientists' thinking and actions, but students should not be expected to personally take on those values. Remind students of this constantly.

Offer learning experiences inside and outside the classroom for both Indigenous and non-Indigenous students. The idea here is to achieve balance. On the one hand, students must learn to be comfortable around a science lab. On the other hand, many teachers find that Indigenous students usually feel at home, highly motivated, and even inspired by natural out-of-doors settings. Consider implementing field experiences whenever possible. Some of these might last less than an hour; others may last several days. These could include short trips to a nearby field or forest, or more substantial visits to fishing camps, trap-lines, sacred sites, and other places of significance where students can enjoy sustained contact with the natural environment. Allow enough time for students to engage in quiet and persistent exploration, and to experience the magnitude and interconnectedness of all life, to which students intimately belong. Instill the ethic of no footprint left behind. Rather than collecting objects, consider taking photos of them. Just as we would want a visitor to respect the objects belonging to our home, we should respect a setting in Mother Nature the same way.

Funding to purchase or rent resources for out-of-doors instruction will be required for such items as canoes, sleds, boats, kayaks, snowshoes, trapping/hunting/fishing supplies, building materials, fuel, first-aid kits, and so on. Many Indigenous students live in poverty-ridden circumstances and are unable to access the land without these vital school activities. Instruction in the out-of-doors as well as in a lab affords more one-on-one interaction between a student and teacher. Complement lab instruction with peer mentoring strategies, group and individual tutorials, group study sessions, and computer aids.

Speaking and listening skills are very important to the oral tradition of passing Indigenous knowledge and wisdom from generation to generation. Therefore, use small-group activities and projects in which students can freely participate, create, develop, and discuss.

Note that 'learner-centred' instruction has two very different meanings. In a Eurocentric context, it focuses on individual students and their needs. In an Indigenous context, however, it focuses on community-centred instruction in which students give back to their community as described above in the sections "Community Contexts" and "Classroom Environment."

Organize an Indigenous knowledge fair exclusively centred on Indigenous cultures and, in particular, their knowledge, wisdom, environmental ethics, perspectives, technologies, and creative innovations. Projects that combine Indigenous knowledge with Eurocentric sciences can also be successful. For example, the "First Nations and Métis Provincial Science Fair" inaugurated by the Federation of Saskatchewan Indian Nations (FSIN, 2009) is held each year in March. To plan such an event, bring together a small advisory group to oversee the entire affair, from proposal to implementation. Involve Elders in suitable ways such as opening and closing the event with a prayer or conducting a pipe ceremony during the event at an appropriately chosen time, but do not ask Elders to be formal judges. Instead, invite them to interact with students as they wish. Acknowledge, honour, reward, and showcase emerging young scientific thinkers who can 'walk in both worlds' effectively. Invite industries, businesses, and other organizations for support. Ask them to sponsor different parts of the Indigenous knowledge fair such as speakers, lunches, materials, resources, electronic equipment, and accommodations. Consider entering a provincial or state 'Indigenous science fair.' Such fairs already exist and a national fair in Canada may soon be a reality.

Deborah's Story Revisited

As described in Chapter 7, Deborah's biology instructors refused to discuss Diné (Navajo) views of Creation, and offered her no other way of resolving her worldview clashes. "Deborah was not asking to remove evolution from the ... curriculum. She was asking her professor for guidance" on how both views might coexist in her mind (Brandt, 2008, p. 839). By refusing her request, her instructors exacerbated Deborah's difficulty in moving back and forth between her Diné culture and the culture of Eurocentric sciences. She thought the professors expected her to abandon her Diné worldview. Deborah needed an instructor who appreciated the cultural challenges she faced daily and who would help her cross back and forth between the two cultures more smoothly and with less psychological risk, the way non-Indigenous science-proficient students do when they move between their home culture and the culture of school science (Aikenhead, 2006).

Brandt (2008) pointed out how cultural borders around Eurocentric science serve instructors, providing them with "a position of legitimacy," "a location of credibility," and "a place of power" (p. 838). As a result, instructors build a moat around their professional identity as science professors without thinking about building bridges across that moat. Deborah's instructors likely held "the widespread belief that scientific knowledge transcends culture [i.e., a belief in universalism and positivism] and therefore science teachers do not need to" concern themselves with matters of culture (p. 839).

Most importantly, a science teacher needs to be open and accepting of students' worldviews and experiences. Instead of seeing these as a liability or disadvantage to learning school science, known as the *deficit model* of teaching, teachers can tap into the holistic and experiential resources of students and treat these resources as assets for academic success. This was the case for Luke, a First Nations boy in Grade 6 who studied the Canadian seashore:

> Clearly, after instruction, Luke continued to have many ideas and beliefs about seashore relationships consistent with a spiritual [Indigenous] view of the seashore and many ideas and beliefs consistent with an ecological view of the seashore [gained from his school science instruction].... It is possible to increase a student's knowledge of science concepts without altering substantially his or her preferred orientation [a spiritual worldview]. (Snively, 1990, pp. 53–54)

Indigenous students can learn Eurocentric science without necessarily being assimilated into its culture and losing their Indigenous self-identity. But, to make this happen, the instruction must be cross-cultural in nature and culturally responsive, as it was for Luke.

By having conversations with Indigenous students who are ready to share their own cultural clashes and ideas for resolving them, a teacher can exchange ideas, forge relationships, and build cultural bridges. These personal conversations are like a camping spot where two cultures come together in a shared neutral space to interact respectfully and with humility.

A culturally responsive science teacher nurtures 'walking in both worlds' or following a 'two-way approach' to teaching—Indigenous and Eurocentric (Battiste, 2000, p. 202; Cajete, 1999). Similarly, in Canada's Mi'kmaw Nation, some Elders talk about "two-eyed seeing" that uses the strengths of both knowledge systems (Hatcher et al., 2009a–b; Marshall, 2007). By walking in both worlds, following a two-way approach, or using two-eyed seeing in a predominantly Eurocentric world, Indigenous students like Deborah and Luke gain the ability essential for accessing social power and economic well-being. Deborah and Luke can become renewable resource managers, scientists, engineers, lab technicians, nurses, or doctors, while simultaneously maintaining their roots in an Indigenous wisdom tradition. In fact, a strong Indigenous identity enables Indigenous students to succeed in school science and science-related programs at university (Hunt & Harrington, 2008).

Student Assessment

Examinations are a major component of school culture, yet they represent a form of communication foreign to many students who have not been explicitly taught how to approach an exam. Such students need help developing effective ways to prepare for and write a test, and they need time to practise those ways throughout the course. Consider offering a practice exam near the beginning of the school year to familiarize students with your tests and how you mark them.

Before a final exam, conduct a series of mini review sessions over a two- or three-week period. Highlight and reinforce important concepts to understand and remember, as well as important skills such as critical thinking, that you expect them to display. Motivate students to ask penetrating questions.

Time limitations on final exams create a challenge for most Indigenous students, as they did for Deborah. If examinations are events for students to communicate what they have learned, then equitable treatment and valid assessment mean that all students should be allowed the time required to do their best. Time limits can marginalize certain students.

When using quizzes and tests, consider organizing them in two parts. First, the usual closed-book questions, and then second, open-book questions, which allow students to use resources such as their textbook, perhaps self-generated review sheets, and even their class notes. Open-book testing rewards in-depth understanding as well as intuitive and critical thinking. It gives students a chance to convey what they really understand beyond memorizing and, in some cases, beyond what a teacher expected them to learn. On the negative side, open-book test items can take more time to mark. They often require more time for students to answer. Overall, the goal is balance. A test could have nine closed-book questions and one open-book question.

Give students plenty of practice answering open-book types of questions before they appear on a test. Students whose past academic success in science relied on memorization may require encouragement to work through their resistance to changes in assessment routines and expectations. Our experience suggests that students must discover that answering an open-book question does *not* entail a whirlwind tour through their textbook or review notes simply to locate the correct answer. There are often several reasonable answers to a good open-book question, reflecting students' creativity and diverse thinking. Such answers are never found directly in a textbook or in a set of review notes. Open-book answers are the result of thinking, not locating. Some well-prepared students do not even open their book or notes to answer an open-book question. The real benefit of open-book questions is in the way students study for a test—they practise *thinking with* what they have learned.

Throughout the semester, use a variety of assessment techniques. Rubrics are helpful because they spell out what is expected, and students can seek clarification of rubric phrases ahead of time. Portfolios can be very productive and beneficial, though time-consuming to initiate. Performance assessment on authentic activities or projects is also helpful. Some teachers use student self-assessments with favourable results.

The type of assessment most conducive to helping students learn is continuous informal feedback—'formative assessment' (Bell & Cowie, 2001). Formative assessment records, such as teacher notes, checklists, and rubrics, can accumulate evidence of accomplishments and can identify areas requiring remediation.

When assessing students' understanding of Indigenous knowledge, it is inappropriate to ask an Elder to evaluate students' performance. But science teachers can evaluate the degree to which a student has demonstrated his or her *engagement* in coming to know. For example, did a student

- indicate a sense that a journey was taken to become wiser?
- demonstrate harmony with nature and other values of the local Indigenous community?
- benefit the community in some way?
- embrace physical, mental, emotional, and spiritual aspects in their in-depth understanding?
- show creativity, originality, and resourcefulness?
- indicate that relationships were formed or strengthened, that responsibilities were taken on, and that protocols were properly followed?

The above list is derived from a judge's rubric for a First Nations and Métis traditional science fair (FSIN, 2009). It illustrates community-based notions of educational success that inspire effective teaching practices, in contrast to conventional notions of educational success that can inhibit effective teaching (Lewthwaite & McMillan, 2010).

Barnhardt and Kawagley found the Eurocentric approach to testing to be very limited. "Such an approach does not address whether that person is actually capable of putting that knowledge into practice" whereas students' coming to know is traditionally "tested in a real-world context" (2005, p. 11).

This higher standard of assessment corresponds to a principal goal of a major international science assessment by the Organisation for Economic Co-operation and Development (OECD, 2006a).

Compared with testing in a real-world context, it is much easier to assess students' awareness or understanding of certain features of Indigenous knowledge studied in a science course. For example, one can assess the depth and breadth of students' explanations of how certain ideas, objects, or events reflect the local collective Indigenous worldview or ways of living in nature. Provide opportunities for students to compare, on specific points, Indigenous and scientific ways of knowing nature.

For group work, consider assessing two aspects: first, the product that a group of students produces, such as a report or skit, and second, the idiosyncratic contribution of each student to the group product. This approach more authentically reflects the assessment of professional scientists who often work in teams and are judged on the merit of their contribution to the team's publication or patent.

Collect as much assessment evidence as feasible over a semester. Invite students to contribute to that collection on their own initiative, for example, by using portfolio techniques. If they choose to submit an item of their work, have them write or orally explain how their item meets an expectation of the course. At the end of the term when all the assessment evidence is in, make a holistic judgment on what mark reflects a student's accomplishments, while at the same time being as consistently equitable as possible for all students.

Indigenous Languages

As demonstrated by Deborah's story, Indigenous languages are inseparable from students' experiences in school science. Teachers can motivate students by giving a modicum of attention to students' Indigenous language, as well as non-Indigenous languages in multicultural settings. Indigenous words and their concepts contain rich sources of hidden knowledge about the natural world (McKinley, 2005; Michell, 2007).

Learn as much of the local Indigenous language as feasible, especially terms and expressions applicable to science teaching, the weather, and safety precautions on the land. A little goes a long way toward opening a window into an Indigenous worldview (Michell, 2007). Do not hesitate to ask students to share with you an Indigenous word or phrase. Greeting local people in their own language is advantageous because it helps forge relationships in an Indigenous community. Remember that variations in spelling and pronunciation among communities in the same language group reflect the natural constant flux of an Indigenous world, as Elder Little Bear (2000) described (see Chapter 6).

In addition to drawing on Indigenous students in class, valuable resources for learning Indigenous words and expressions include Elders and language instructors, as well as personal experiences with Indigenous speakers on the land. When an Indigenous word does not seem to translate accurately, Elders can sometimes identify a common understanding by getting a person to experience what the Indigenous word means, so it does not lose its meaning when translated into English or French.

Encourage students for whom English or French is a second language, to work out the meanings of scientific ideas in their first language and to translate them back into English or French, a process called 'back translation.' This was the process that led us to replace "knowledge" with "ways of living" (see Chapter 5). Back translation can help a teacher better understand students' worldviews, and it may motivate students to understand Eurocentric sciences better.

If support resources are available, give students opportunities to use their Indigenous language in assignments related to observations, hypotheses, experimental results, and the development of theories. Indigenous languages and place-based dialects embody detailed concepts that have been used for thousands of years to name and classify the natural world. The Office of the Treaty Commissioner (2009) describes many strategies and activities that foster an Indigenous language in a school.

If possible, promote the use of Indigenous language concepts to identify and classify animals and plants alongside scientific concepts and classification schemes. The left-hand page in a student's notebook could be devoted to Indigenous ways, while the right-hand page could represent scientific ways (Aikenhead, 2002). This format reinforces a balance of coexisting perspectives. For instance, on field trips or excursions to hunting sites where moose have been killed, have an Indigenous speaker identify and document parts of the animals using a native language and concepts (Michell, 2007). Challenge students to write on the right-hand page in their notebooks any applicable scientific anatomical and physiological information. Provide assistance as required. Content on both pages (Indigenous knowledge and Eurocentric) should be open to student assessment. A similar approach may be possible with research reports assigned to students—bonus marks for reports that include instances of the two-page (Indigenous/Scientific) format.

It is important not to assume that students know terminology in the language of their traditional territory. Many First Nations languages have been systemically discouraged and some have been entirely eradicated. At times, there is also reluctance to re-learn an Indigenous language because it is regarded as impractical in current times. It may also be resisted at home by adults scarred by residential schooling. However, by introducing some Indigenous words or phrases in the science classroom, teachers can have a significant impact on the formation of a stronger Indigenous self-identity, which often results in greater academic achievement. Wagamese (2008) recounts the first time he spoke out loud the Ojibwa word *Peendigaen* (come in). "I felt as if I'd truly spoken for the first time in my life" (p. 137).

There is a subtle underlying sequence within some of the ideas suggested above: Indigenous knowledge comes first, scientific knowledge second. This sequence is important because the validity of local knowledge is first established. Then scientific knowledge is introduced as another culture's valid way of making sense out of the event. Eurocentric science often enriches or clarifies one small aspect of the Indigenous knowledge. In the "Snowshoes" unit summarized in Appendix A, for instance, the concept of pressure enriches a portion of the *physical* aspects of understanding snowshoe technology, but the concept of pressure has nothing to say about the *mental*, *emotional*, and *spiritual* aspects of understanding snowshoes. Therefore, Eurocentric science neither replaces nor corrects the Indigenous knowledge—the two coexist similar to the way yin and yang coexist in Eastern philosophies.

A capable science teacher augments students' reading abilities. When giving students a reading assignment in handouts, books, or articles, skim and scan the reading with them in advance. Introduce new vocabulary. Remind students to make an educated guess about an unfamiliar word's meaning by inferring a meaning from the context. Highlight key information in the text so students will know what they should concentrate on. Develop questions for students to answer while reading. These questions can guide students about material that they should emphasize and reflect on. Questions could require students to make connections to previous experiences and knowledge they have acquired. Students' answers can form the basis of a good class discussion.

When posing a question in class, allow enough time for Indigenous language speakers to offer a verbal response. They need time for second language processing—to interpret the question from English into their own language and back into English to respond. Do not interrupt the process or allow others to interrupt. Listen to students' perspectives with a sense of compassion and patience. This is all part of culturally responsive teaching. Of course, this advice also applies to classrooms in which the conventional language of instruction is not English.

Become aware of non-verbal patterns of communication specific to the classroom's Indigenous community by asking people about them. For example, it is common for some teachers to expect eye contact when teaching a class or speaking individually with a student. However, many cultures regard eye contact as confrontational and at odds with their socialization related to harmony and balance. These more traditional students may be showing a teacher respect by not making eye contact. But they may also be expressing the power imbalance between a teacher associated with colonizers and students identified with the colonized. In other words, the visual non-contact could be shaped by historical and contemporary events of oppression. Is it a cultural response or a power-imbalance response? The question is difficult to answer but necessary to consider if respect for students' cultural practices is prominent.

Teacher Expectations

A teacher's expectations are a major determinant of Indigenous students' achievement in school science (Michell, 2007; Taylor et al., 2009). It does not help students to set lower standards. Instead, a teacher can take Indigenous students on a different pathway to reach the teacher's high expectations. For instance, to prepare students to produce a report or to write an important exam, a teacher could explicitly brief them on how to meet expectations in these academic contexts, and offer them chances to practise. But, in the end, the same standard of academic achievement is expected of all students.

Some Indigenous adults attribute their employment success to a teacher who did not lower their professional standards or ignore students' potential, but who offered a humanistic balance between culturally responsive instruction and clear expectations for academic success (Wagamese, 2008). A teacher should set this tone on the first day of class, and let it pervade the classroom environment.

Although student peer pressure in some urban settings may discourage high expectations, a science teacher will likely find support within the urban community itself: "Not only do urban Aboriginal peoples see higher education as a path to a good job or career for their own generation, many say that they hope higher levels of education will be key to how future generations of Aboriginal peoples will distinguish themselves from their ancestors" (Environics Institute, 2010, p. 3).

Acknowledge student successes and achievement by immediate and consistent laudatory feedback. Give praise that is specific, no matter how small. Students need to know they are on the right track with their thoughts and that they are moving along their learning trajectory. Some Indigenous students may prefer praise in private so they do not appear superior to their peers. Time, as well as venues outside of class, are needed for teacher-student interaction in which students can discuss assignments, lessons, and problems. Non-academic events at school, such as a fun 'chemistry is cooking' lesson, held either in the classroom or during extracurricular activities, provide much needed social interaction between a teacher and students. However, most importantly, a teacher must show students that he or she is approachable and a good listener.

Do not expect an Indigenous student to represent their Indigenous nation or to be an encyclopaedia of Indigenous knowledge. Indigenous ways of knowing are acquired over a lifetime, which includes the time students spend in a science course enhanced with Indigenous knowledge. Many Indigenous students and their families have not had the chance to learn much Indigenous knowledge from their life experience due to negative factors such as the legacy of residential schools. Therefore, an Indigenous student may not know the answer to your question. For those whose traditional lifestyles have been preserved, some knowledge is sacred and private, and will not be shared openly with just anyone. Consult privately with students before asking them to share cultural teachings, rather than putting them on the spot in front of their peers.

Mr. Chang's Story

In his first year of teaching, Michael Chang taught Grade 9 biology in a California city school to immigrant children, some of whom shared his Hmong Indigenous culture. The Hmong Nation of Laos underwent ethnic cleansing as a result of repeated attempts at genocide, mostly during the 1960s and 1970s. Like many Indigenous nations, the Hmong people fear cultural genocide. Mr. Chang was a child when his family fled Laos. The following story is abstracted from a case study by Chang and Rosiek

(2003). It is the only published detailed case study concerning teachers in cross-culture science education. In this chapter, Mr. Chang's story focuses on his reflections about one key incident—a challenging question posed by one of his students about Eurocentric science and her Hmong Indigenous culture.

Mr. Chang had no desire to return to believing in everything his grandparents once did, but he regretted what he had lost of his Hmong culture during his life in the United States. His professional goal was to teach Eurocentric science to *all* students, so they could decide what to believe based on rationality, not on authoritative coercion he associated with his parents.

A cellular biology unit was the first unit taught in Grade 9. Mr. Chang's final examination for the unit addressed all the content expected of him as the instructor. What was unexpected was the choice he gave his Hmong students of answering in Hmong either orally or in writing, because he was interested in their meaningful learning.

As students finished his exam, they would drop it on his desk and then leave the room. He was surprised that Lia, a most capable Hmong student, was the last to hand in her test.

When she did so, Mr. Chang asked, "Well, Lia, are you finished with your first high school science test?"

"Yes, I think I did them right," she replied.

"Yes," he assured her, "I am sure you did."

"I have a question for you, Mr. Chang."

"Yes, Lia?"

"You don't really believe this do you, Mr. Chang?"

"Believe what, Lia?"

Pointing to her test paper, she clarified, "In the 'cancer.' I mean you are Hmong. You know a person gets sick when they lose their *pleng* [soul], and they need the *twix neeb* [shaman] to help guide their *pleng* back so they can be good again."

Mr. Chang thought to himself, "Her question is not really a question at all, but a statement insisting that I recognize and validate the traditional Hmong explanation of sickness. I want to teach [Eurocentric] science, but I don't want to condemn traditional Hmong culture. I want to keep the two separated. Lia's question is pitting them against one another in a way I have been trying very hard to avoid. Hmong beliefs about diseases are connected to other Hmong beliefs about the world. This fabric of beliefs brings people together and enables them to live and love and support one another."

A few years earlier, his mother had been diagnosed with breast cancer and an experimental treatment was kindly offered to her without charge by the hospital because the family had no health insurance. But his father refused to have his wife 'experimented on,' and he insisted she go to a Hmong shaman (*twix neeb*). Mr. Chang and his brother unsuccessfully tried to intercede in the shaman's treatment, but to no avail. However, after the shaman's treatment, their mother went into remission, as confirmed by local medical doctors.

"Up until now," thought Mr. Chang, "I have resolved the tension between these two views by oversimplifying things. I let myself believe that cultural responsiveness in my science classroom would just be a matter of respectfully acknowledging folk understandings—insisting they were separate issues. Lia's unexpected question, and the concern it makes me feel for her, are making me see how much more complicated it is than that. Her question suggests that scientific understanding can't be sealed off from the rest of students' lives, like a specimen in a jar, just to be observed in a specific classroom lesson."

Mr. Chang finally answered Lia. "I don't want to answer such an important question too hastily. Let's talk about this again after class on Monday."

"Okay, Mr. Chang. Thank you," she cheerfully replied.

That weekend, Mr. Chang had much to ponder. "My experience tells me that in order to deal with cultural differences in the science classroom, teachers need to know more than their subject matter and pedagogy. They need to understand something about the specific culture of each of their students and how it relates to the cultural assumptions in the curriculum they are teaching. I think, like most science teachers, I have been taking the status and authority of [Eurocentric] science for granted. I wanted to hide behind the idea that science is just about the physical world. I am beginning to see that this attitude is irresponsible. Science is not experienced by my students as something separate from their questions about religion, identity, and community commitment. Even if a teacher philosophically believes that science and culture should be separated in our thought, that doesn't change the way they actually operate in our society and the way children experience them. As teachers, we need to deal with real children and their actual experience of science, not ideal children and ideal conceptions of science subject matter."

As Mr. Chang thought about what Lia needed to hear, he drew on what he knew of Hmong culture and history. He concluded, "If Lia ever finds herself in a situation similar to the one my mother faced, I want her to be able to take care of her body better than my mother knew how to."

When Lia walked into his classroom Monday afternoon, Mr. Chang assured her, "I have thought a lot about your question." Then he asked to speak with her after class.

When that moment arrived, Mr. Chang affirmed to Lia both his belief that cellular biology explains cancers and his belief that the Hmong community's values and rituals are important to Hmong people's health. He added, "There are many people who criticize Hmong traditions and beliefs, but I do not think that is a reason to abandon them."

After a long pause, Mr. Chang spoke again. "The Hmong are changing in America, aren't we?"

Lia nodded.

Then he asked her, "What changes do you see?"

"Children are less respectful of their parents. Like my brother. He made my mother cry. He stayed out all night last week."

"I'm sorry to hear that. That is bad. Do you see any good changes?"

After a few moments of thought, she replied, "In America I get to go to school longer."

"Yes, it is good for you to get more education. We will have to work hard to make the changes more of the good kind and less of the bad kind."

"My brother says the *twix neeb* is crazy," Lia interjected.

Identifying with her brother for a moment, Mr. Chang asked, "Why does he say this, Lia?" She gave an explanation of the tensions and strife in her family over the shaman as opposed to American doctors.

Not wanting his scientific belief in cellular biology to be seen as an endorsement of her brother's erosion of family bonds, Mr. Chang said, "I am sorry your brother is making your mother cry. I can talk to him if you want me to." Lia agreed but was not hopeful of the results.

Then Mr. Chang talked about his own personal conflict between Eurocentric science and Hmong culture, and about his mother. He added, "Medicine is hard for our parents. They have reason to be afraid of doctors, hospitals, and government agencies. In the past, such agencies have done very bad things to the Hmong." Lia nodded emphatically in agreement. Mr. Chang continued, "But going to the medical doctors when you are sick is the right thing to do here in America. I don't believe hospitals here want to experiment on Hmong people, but I do believe hospitals are ignorant of Hmong tradition and that most government agencies have too little respect for the Hmong. I think the Hmong in America need both the medical science and the Hmong traditions to live well. It isn't easy, and I'm still learning how to do it." He ended their conversation with, "Keep asking me these questions and we can think about it together."

Lia left, but not before thanking Mr. Chang and asking how his mother was feeling.

Mr. Chang could not decide what Lia made of their discussion. But he had much to contemplate. "Being a good science teacher," he thought, "requires a more complicated view of things. Teaching [Eurocentric] science fully and well, and in a way that avoids blatant colonialism, requires some knowledge of the student's cultural community and the history of that community. This cultural knowledge doesn't have to be comprehensive, nor does a teacher have to be a cultural insider to have it. But it is essential to understand some of the stories students bring to the classroom. It is only through this knowledge that teachers can avoid falling into the trap of easy and uncharitable oversimplifications of [Indigenous] experience, which in turn inhibit the truly responsive science teaching that reaches *all* students."

The details in Mr. Chang's story, set in a multicultural inner-city classroom, may differ from other school settings, but his experience and reflections apply to most Indigenous cross-cultural school science classrooms that strive for culturally responsive teaching.

Conclusion

Sutherland and Henning (2009) developed a framework for Indigenous science education that guides Indigenous communities in developing or assessing their local science curricula and school programming. This framework for lifelong learning in Indigenous settings stipulates essential elements for successful, culturally responsive, school science. Two sets of elements are integrated in their framework: (1) the importance of Elders or Knowledge Keepers, culture, language, and experiential learning; and (2) coming to know, culturally relevant science teaching, social and ecological justice, and ecological literacy. This framework provides an invaluable overview for bridging Indigenous and scientific ways of knowing nature.

Also very helpful are the effective classroom practices for Inuit students in Grades 5 to 8, identified by Lewthwaite and McMillan's (2010) in-depth research with students and teachers in Nunavut, Canada, and corroborated by Mäori experiences in Aotearoa New Zealand. This research produced 10 characteristics of successful cross-cultural science teaching that support this chapter's advice on how to build bridges of understanding for culturally responsive school science (CRYSTAL, 2010).

The Australian Academy of Science's (2009) website describes characteristics of "quality teaching and learning" for Indigenous students. The Academy emphasizes (1) teacher-student and teacher-community relationships, (2) a supportive classroom environment for Indigenous students, (3) high expectations of students based on individual learning achievements, (4) learning styles, and (5) links with Indigenous communities. The Academy's descriptions summarize some of the detailed advice offered in this chapter. Their website also has links to videos and other resources.

There is a persistent message in Indigenous-based, cross-cultural science education worldwide—connect all students, Indigenous and others, with the land and environment. Elders, Knowledge Keepers, community members, parents, teachers, and many scientists would agree there is a need to rethink and re-identify the world in which we live.

> We need to help our children develop a relationship to the land to make them aware of how humans affect Mother Earth. They grow disconnected from the land. (Traditional Knowledge Keeper Judy Bear, quoted in Johanson et al., 2009, p. 41)

As human beings, students require opportunities to explore their place within the natural world. Indigenous perspectives on the natural world are still very much an untapped source of community-based knowledge and wisdom-in-action that can enhance a person's core of interconnectedness and well-being.

It is worth repeating that Indigenous students' backgrounds should be treated as a resource to draw upon for academic success, rather than a deficit to be overcome in learning school science. With greater understanding of Indigenous knowledge, a teacher like Mr. Chang has more ways to recognize students' cultural resources. This is a science teacher's journey of lifelong learning.

Young people should be engaged in school science experiences where they feel comfortable exploring who they are—as a person of the planet, a person who inhabits Mother Earth, or a hybrid of both. All young people must work toward their obligations and responsibilities for living in a sustainable manner. They will become more literate in knowing nature.

The best way to engage Indigenous students, as Mr. Chang did, is for science teachers to begin to build cultural bridges between their familiar scientific world and an Indigenous world of students. Science teachers must experience for themselves Indigenous culture, language, and cross-cultural science teaching. Science instructor Andrea Belczewski (2009) did this and wrote her story for others to read. We highly recommend it. Knowledge of teaching, like authentic Indigenous knowledge, requires experiential learning.

In the Plains Cree language, "bridge" is *Asokan*, and *Ni Asokanihkan* means "I make a bridge." Through a teacher's personal effort to build cultural bridges, he or she will motivate students toward interconnectedness, well-being, and academic success. Greg Cajete (1999) expressed this idea as follows, referring to "science" in its pluralist sense as we do in this book, thus encompassing Indigenous knowledge, Eurocentric sciences, and other cultural ways of knowing nature:

> The construction of a bridge between [Eurocentric and Native American] mindsets concerning the natural world affords the student of science a viewpoint and orientation which allows for a broader and more realistic perspective of science as a whole process. In doing so, it allows students a greater opportunity to develop an appreciation of science as a highly flexible and creative tool for understanding the natural world as well as their own relationship to that world. (p. 182)

We encourage science teachers and those preparing to be science teachers to contemplate and select ideas from this book to begin their journey into building cultural bridges, strengthened by more ideas and insights gained through experiences with Indigenous students, community people, and their place-based wisdom. Bridging cultures is an indispensable step toward cross-cultural or multicultural teaching of an enhanced science curriculum in a culturally responsive way.

APPENDIX A
A Cross-Cultural Science Unit

You may be curious about what a cross-cultural science unit might look like. The following overview describes the unit "Snowshoes," developed by Knowledge Keepers in Île-à-la-Crosse, Saskatchewan, and teacher David Gold. This unit appears in the *Rekindling Traditions* website (Aikenhead, 2000).

The unit begins in the domain of local Indigenous knowledge. Indigenous and non-Indigenous students make a personal connection to nature in the first lesson with an afternoon of snowshoeing. The Indigenous key value of happiness is given prominence in this lesson. In a debriefing session, students hear how their community is rich in knowledge about snowshoes. Students will later explore this richness on the Internet and in print materials. To gain access to local knowledge about snowshoes, students learn the protocol for approaching community Knowledge Keepers, as well as the proper way to conduct interviews with them. Interview questions are composed by the class and then used by groups of students in the community. The local knowledge gained by students is shared and synthesized in class. Teacher David Gold becomes a co-learner with his students, modelling an adult's life-long learning. To acquire this Indigenous knowledge, non-Indigenous teachers and students consciously bridge their Euro-Canadian culture and the predominant Métis culture of Île-à-la-Crosse.

After students have gained a good introductory understanding of Indigenous knowledge concerning snowshoes, such as their structures, materials, and fabrication, their teacher consciously brings both Indigenous and non-Indigenous students into the culture of Eurocentric science. (Typically, both an Indigenous home culture and a non-Indigenous home culture differ noticeably from that of Eurocentric science, except in the cases of science-proficient students; Aikenhead, 2006.) Students cross into the culture of Eurocentric science to learn a very different type of understanding of snowshoes. For example, the accepted scientific explanation for how snowshoes stay on top of the snow is *pressure = force/area*. The scientific key value taught in this lesson is that "scientific ideas are abstract, generalizable, and often mathematical."

Teachers are aware of the commonsense preconceptions that students invariably bring to class. For instance, when comparing two quantities, students tend to confuse the mathematical idea of *difference* for the idea of *proportion*. Teachers try to design their lessons so that students can journey into the culture of Eurocentric science as smoothly as possible and with the least amount of risk to their psychological safety. The scientific concept of pressure ($P = F/A$) is contrasted with the commonsense idea of a 'push on the snow' that students experienced by walking on snowshoes. Hands-on activities, problem-based learning, and calculations are completed using an assortment of snowshoes and an array of everyday situations.

Students' personal interests are piqued by researching a topic related to snowshoes and chosen by a student. Meanwhile, students play the role of scientist when they design a classic experiment to observe the ability of various types of snowshoes to handle different snow conditions. This experiment has two independent variables—type of snowshoe and snow conditions—and requires a three-dimensional graph to organize the results. Such a complex experimental design is seldom attempted even in Grade 12 physics, but in this situation the experiment successfully engages Grade 8 students, Indigenous and non-Indigenous. Various technological problems must be resolved; for instance, how to measure the dependent or responding variable (the ease of walking on snow). This challenges students' creativity and ingenuity. The scientific idea of reliable data comes alive during the experiment. The question, "How do you know?" is asked by more than one student during classroom discussions.

The data collection activity results in students going back into nature on snowshoes. But, this time, students play the role of teams of scientists measuring which type of snowshoe is best for which type of snow condition. Variables now have concrete significance for students. They can also feel at ease playing the role of a scientist by talking and thinking like a scientist because they are not required to dismiss their commonsense knowledge, whether Indigenous or non-Indigenous, in the process. For Indigenous students, their cultural self-identities are nurtured.

Students' grasp of local knowledge about snowshoes and their ability to learn Indigenous phrases found in their community—such as *Asâmak,* the Métis word for snowshoes—are assessed, forming a portion of a student's total mark for the unit.

This unit is one among six developed in the project *Rekindling Traditions* (Aikenhead, 2000; http://www.usask.ca/education/ccstu/units/index.html; retrieved Dec. 17, 2010). The five other units are:

- "Nature's Hidden Gifts" (Morris Brizinski, Beauval),
- "Wild Rice" (Gloria Belcourt, Pinehouse Lake),
- "Trapping" (Keith Lemaigre, La Loche),
- "Survival in Our Land" (Earl Stobbe, Timber Bay), and
- "The Night Sky" (Shaun Nagy, La Loche).

The *Rekindling Traditions* website also has a substantial "Teacher Guide." Its table of contents helps the reader locate topics such as "Treating Aboriginal Knowledge with Respect." Another document, "Stories from the Field," describes how teachers gained community support and involvement for developing and teaching their units.

These units are an example of integrating Eurocentric sciences into an Indigenous knowledge context. Another approach is to integrate Indigenous knowledge into a Eurocentric science context in a way that maintains their validity and coexistence. This second approach is found in most school science curricula developed by ministries and departments of education, and in the textbooks that support teachers in implementing those curricula (Pearson Canada, 2010).

APPENDIX B
Questions for Reflection and Discussion

The following chapter-related questions ask readers to reflect on what they have read and to clarify their ideas concerning specific topics. Some questions also invite the reader to contemplate implications for their teaching. Professional development discussion groups could effectively use these questions.

In addition to the many references cited throughout the book, this appendix offers "Targeted Reading" for Chapters 3 to 8 by listing a small number of comprehensive sources with which to explore topics more fully.

Chapter 2
Various justifications exist for Indigenous knowledge being included in school science:

    A. Equity and social justice
    B. Strength of a nation's economy
    C. Improvement of Eurocentric science
    D. Preparation of science-oriented students for science careers
    E. Indigenous sovereignty and cultural survival
    F. Enhancement of human resiliency
    G. Positive results of integration

1. What other justifications, if any, would you add to this list?

2. Which justifications (including your additions) are most compelling to you personally? Explain your choice.

3. Individual students will relate more to one justification than another. Consider a variety of viewpoints held by students, and match each student viewpoint to a justification that would likely make the most sense to that student's viewpoint.

4. Suppose the topic of your school's new and enhanced science curriculum came up during a conversation with a person in your community, such as at a parent-teacher interview or with a neighbour in a grocery

store aisle. Some of the justifications listed, including your own, will make more sense to some people than to others. For each of the seven or more justifications identified earlier, anticipate a person in your community who might relate most strongly to that reason for including Indigenous knowledge.

Chapter 3

1. The history of modern science includes dramatic changes to past institutions: Natural philosophy in 1660, science in 1831, and research and development (R&D) during the twentieth century when science generally became integrated into the social, economic, and political fabric of the twenty-first-century societies we know today. The names given to the practitioners within each institution have changed from "natural philosopher" to "scientist," but there is no new name for the practitioners of R&D.
   a. What reasons might explain R&D practitioners' identification of themselves as scientists?
   b. Suggest characteristics that identify practitioners of R&D (the people involved in "post-academic science") as scientists.

2. This book defines Eurocentric science as "what scientists do." This is an *operational* definition. Many other types of definitions for Eurocentric science exist, such as conceptual, philosophical, stipulative, consensual, and so on. In the past, what definition of Eurocentric science have you found to be useful? What type of definition is it?

3. A Native American scholar defined a generic *pluralist* science as "a story of the world and a practiced way of living it" (Cajete, 2006, p. 248). A Japanese science educator defined it as "a rational perceiving of reality" (Ogawa, 1995, p. 588).
   a. How would *you* define a generic pluralist science?
   b. What is your story or rational perceiving of the world?

4. Although each person's worldview is unique in some ways, it will often have features in common with groups of people.
   a. Take gender as an example. List three common features that you consider to be associated with a predominantly feminine worldview, and likewise for a predominantly masculine worldview.

b. Suggest two or three features of a predominant Eurocentric worldview. This may be challenging for people of Euro-American cultural ancestry because "If you want to understand water, don't ask a fish."

Targeted Reading:
Aikenhead (2006, Ch. 2 & 7), McKinley (2007), and Ziman (2000).

Chapter 4

1. The vast majority of scientists presuppose the world is comprised of material and non-material parts, an idea known as Cartesian duality. Eurocentric sciences (ES) are concerned with only the material world.
   a. Explain how the scientific enterprise is strengthened by restricting itself to the material world.
   b. Explain how this restriction might be a limitation in any way.

2. a. How might the doctrine of universalism help explain the underrepresentation of Indigenous students in university science and engineering programs?
   b. Universalists hold that science is culture-free and therefore they would not call science "Eurocentric science." What facts or evidence would seem to contradict their claim?

3. a. In your prior thinking about scientific evidence and theories, were you more of a realist, a critical realist, an instrumentalist, a social constructivist, or a postmodernist?
   b. How do you account specifically for holding that position?
   c. What are your current concerns or questions, if any, about the nature of scientific evidence and theories?

4. Positivism makes ES look as if they are 'cut and dried' (objectively straightforward) instead of a 'messy' affair complicated, in part, by subjective influences such as cultural presuppositions and social values.
   a. List some advantages of conveying to students an image of a 'cut and dried' Eurocentric science.
   b. List some disadvantages.
   c. In science-related social controversies, such as what causes climate change, the public's understanding of ES may influence their viewpoint. Pick a science-related social controversy and describe two possible viewpoints on that controversy: one from the 'ES are cut and dried' perspective and one from the 'ES are messy' perspective.

d. How do the ideas of *frontier science* and *core science* help clarify the two ways of understanding ES ('messy' and 'cut and dried')?

e. What are some classroom practices that do *not* teach an outdated positivist view of science, but convey the 'messiness' of ES without undermining the smooth management of the classroom?

f. How can a teacher emphasize for students the importance of rigorous, valid, empirical evidence in Eurocentric sciences, even though these sciences are embedded in some subjective influences?

5. ES may be described as value-laden. Suggest ways that three types of values (social, community, and epistemic) could be involved at different stages in the planning, execution, and reporting of a scientific investigation.

Targeted Reading:
Goldstein and Goldstein (1981, Ch. 2–3, & 6–13), Longino (1990), Wong and Hodson (2009, 2010), and Ziman (1984, Ch. 3–16).

Chapter 5

1. In some Cree-English dictionaries, the Nehiyaw (Plains Cree) word *Kiskeyitamowin* is translated as "knowledge," although some dictionaries do not translate it at all. What ideas are lost in this translation?

2. The Greek myth of the Trojan Horse is an excellent metaphor for how certain English or French terms, such as "knowledge" or "connaissance" and "savoir," may continue to intellectually colonize Indigenous people.
   a. What is the story of the Trojan Horse?
   b. Drawing upon the Trojan Horse myth, explain how the use of certain terms may contribute to the intellectual colonization of Indigenous students, however unintended.

3. Consider the expressions "to learn" and "coming to know." Although "to learn" is associated with Anglo cultures, and "coming to know" is associated with Indigenous cultures, *non*-Indigenous people can engage in coming to know within their own culture. Describe what someone of British ancestry, or any other non-Indigenous ancestry, would be doing if they were coming to know something in an Anglo or other cultural context.

Targeted Reading:
Cajete (1999), and McKinley (2007).

Chapter 6

1. Although scientific knowledge offers valid predictions concerning the physical world, it is typically silent on what ethical and wise action should be taken in response to those predictions. Yet, most conventional school science curricula are not limited to scientific knowledge, but move into the realm of ethics and wisdom by promoting values such as stewardship. How is it possible that Indigenous ways of living in nature can deal simultaneously with the physical world, ethics, and wisdom, while scientific knowledge mostly cannot?

2. Eurocentric people talk in terms of looking after the planet by taking responsibility in such ways as modifying plants genetically, conserving energy, recycling, or minimizing pollutants entering the environment. Science curricula call this responsibility *stewardship*. This term in the context of Indigenous cultures may appear to carry a similar connotation that Mother Earth is being cared for. However, an Indigenous meaning of stewardship is different from that found in a Eurocentric context.
    a. Specifically, what Indigenous ideas get lost in translation when the term "stewardship" moves from an Indigenous context to an anthropocentric context of a Eurocentric science curriculum?
    b. What Indigenous expression or symbol is often used to reinforce an Indigenous meaning of stewardship?
    c. Specifically, how might a comparison of the two meanings of stewardship be introduced to older students in Grades 8 to 12?
    d. To make a science class interesting, a teacher may illustrate a scientific concept by describing a commercial process found in a local resource extraction industry. Some Indigenous students, however, may feel emotionally uncomfortable with this description. Identify a likely source of their discomfort.

3. Indigenous worldviews are generally holistic and relational, among other attributes. Identify a group within a *non*-Indigenous culture who is also associated with holistic or relational thinking.

4. Is the difference between knowing "how the universe works" and "what the universe is" similar to the difference between knowing "how your body works" and "who you are as a person"? Explain.

5. a. Explain how rationality can be thought of as culturally pluralistic.
   b. The boundary between rational and irrational is a grey area. Describe an unambiguous case of rationality and of irrationality to express what the two terms mean to you.

6. a. Clarify a distinction between holism and monism.
   b. In what ways might Indigenous spirituality relate to each?

7. Summarize in your own words what Indigenous knowledge means, based on the Indigenous scholars you have read or heard.

8. What meaning do you make of the following statement? "Bridging cultures can be thought of as a moral or ethical act, as well as an intellectual achievement."

Targeted Reading:
Cajete (1999), Knudtson and Suzuki (1992), MacIvor (1995), Michell (2007), Peat (1994), and any book in Appendix D.

Chapter 7

1. Think of an Indigenous or non-Indigenous student you know who is similar to Deborah in some way. How might that student cope with a culture clash in school science?

2. As part of an assessment exercise, a teacher can compose a question or a set of directions that clearly conveys to students that the goal is *understanding* something, rather than *believing* it. Demonstrate the difference by wording a test question in those two different ways.

3. a. How can a cognitive process, such as observing, be culture-based?
   b. How can a cognitive process, such as interpreting, be paradigm-based?

4. What would you change in Tables 7.1 and 7.2 to improve them?

5. From your conversations with Indigenous students who have shared with you their own culture clashes and ideas for resolving them, compose a vignette like Deborah's story. These vignettes should be shared with other science teachers and published in professional journals or bulletins for all to read.

Targeted Reading:
Barnhardt (2006), Cajete (1999), and the Canadian Council on Learning (2007a).

Chapter 8

1. Each teacher will have his or her own way of creating a culturally responsive classroom to accommodate Indigenous students. List 10 ideas that would help a teacher accomplish this goal.

2. The culture of school science has not normally encouraged teachers to nurture a culturally responsive, cross-cultural classroom environment.
   a. To support science teachers in this new endeavour, what policies need to be in place at the school level and at a higher level of educational authority?
   b. Reflect on the ideologies that tend to underpin conventional science teaching, and describe how school science might be modified by changing some of its ideologies to accommodate Indigenous students?

3. To teach about Indigenous knowledge, one must acknowledge the balance of the mental, spiritual, emotional, and physical in all events. In this context, how might Indigenous spirituality be introduced in a science classroom in a way that is respectful to all students?

4. How might a teacher collaborate with out-of-school people to help improve both Eurocentric science and Indigenous knowledge literacies for both Indigenous and non-Indigenous students, that is, to improve students' literacy in knowing nature?

5. Using the listing of 10 recurrent learning strengths found among Indigenous students (the section "Indigenous Student Learning"), produce your own personal profile of recurrent learning strengths ("__ more than __"). If you do not consider yourself favouring one over the other on any item, then write "balanced between __ and __."

6. What teaching methods could help incorporate Indigenous knowledge into sustainable ways of thinking and being?

7. How might you engage in professional teacher research in your classroom ('action research') to develop cross-cultural science lessons?

8. In what ways, if any, would your reaction to Lia's question have differed from Mr. Chang's reaction?

Targeted Reading:
Aikenhead (2000, "Teacher Guide" and "Stories from the Field"), McKinley (2005), Michell (2007), and Appendix C.

APPENDIX C
Website Resources

In addition to the websites cited in the reference section, the following resources may be of interest to science teachers and students. These entries are primarily organized by country of origin. Within each country category, the sites are listed alphabetically based on a boldfaced key word in the organization, institution, or website title. These sites are only a fraction of what is on the Internet, of course. Readers should be aware that website content is not necessarily constant and URL links are not always active. Teachers should not share with students any links that they have not personally verified.

CANADA

Canadian **Aboriginal Science & Technology** Society
CASTS is a national, non-profit corporation with the purpose of increasing the number of Indigenous people in S&T occupations in Canada. It supports teachers by holding conferences.
http://www.casts.ca/. Retrieved Sept. 14, 2010.

Integrating **Aboriginal Perspectives** into Curricula
Produced by Manitoba Education and Youth in 2003, this manual is a resource for curriculum developers, teachers, and administrators.
http://www.edu.gov.mb.ca/k12/docs/policy/abpersp/ab_persp.pdf. Retrieved Sept. 14, 2010.

First Nations **Films**
This is a commercial source of Indigenous documentary films and videos.
http://www.firstnationsfilms.com/catalogue.htm. Retrieved Sept. 14, 2010.

Assembly of **First Nations**
Their website updates the reader on a wide variety of programs, projects, information, and news.
http://www.afn.ca. Retrieved Sept. 14, 2010.

This website includes "History of Indian Residential Schools," an interactive unit to learn the history of residential schools anywhere in Canada. This page can be accessed directly at http://www.afn.ca/residentialschools/history.html. Retrieved Sept. 14, 2010.

Indian and Northern Affairs Canada
This Government of Canada site provides information related to Indigenous education.
http://www.ainc-inac.gc.ca/index-eng.asp. Retrieved Sept. 14, 2010.

This website includes "Kids' Stop," a fun zone for children with lots of interesting facts about Indigenous history and languages, educational games and stories, and additional classroom resources for teachers.
http://www.ainc-inac.gc.ca/ach/lr/ks/index-eng.asp. Retrieved Sept. 14, 2010.

Saskatchewan Indian Culture Centre
The Saskatchewan Indian Culture Centre is the source for the teaching unit *Practising the Law of Circular Interaction*, an Indigenous ecology curriculum that parallels a Eurocentric science ecology curriculum. Videotapes are available. The section "Principles and Values of Saskatchewan First Nations," produced by Elders, is particularly useful.
http://www.sicc.sk.ca/. Retrieved Sept. 14, 2010.

Indigenous Studies Portal Research Tool
See the "Science & Technology" and "Indigenous Knowledge" pages at this digital library.
http://iportal.usask.ca/. Retrieved Sept. 14, 2010.

Inuit Tapiirit Kanatami
ITK is the national voice of Canadian Inuit; it provides information and publications.
http://www.itk.ca/about-itk. Retrieved Sept. 14, 2010.

Department of Justice, Canada
This webpage contains The Constitution Act, 1982, Part II, in which the rights of Indigenous peoples of Canada are found.
http://laws.justice.gc.ca/en/const/9.html#anchorsc:7-bo-ga:l_II. Retrieved Sept. 14, 2010.

Canadian Council on Learning
CCL's Aboriginal Learning Knowledge Centre was created to provide a collaborative national forum that would support the development of effective solutions for the challenges faced by First Nations, Métis, and Inuit learners. Documents can be downloaded from http://www.ccl-cca.ca/CCL/AboutCCL/KnowledgeCentres/AboriginalLearning/index.html.
Retrieved Sept. 14, 2010.

For example, three holistic learning models (First Nations, Inuit, and Métis) can be found at http://www.ccl-cca.ca/ccl/Reports/RedefiningSuccessInAboriginalLearning.html. Retrieved Sept. 14, 2010.

National Métis Council
The website keeps citizens throughout the Métis Nation informed on developments and initiatives, and is a leading source of information about the Métis Nation.
http://www.metisnation.ca/office.html. Retrieved Sept. 14, 2010.

Native Access to Engineering Program
Teacher professional development workshops are held at biannual conferences (*Dream Catching*) to support the integration of Indigenous knowledge into school science and mathematics, particularly for younger grades. The program has produced teaching materials since 1998. Each of its 25 different topics has a newsletter, worksheet, and teacher's guide. One part of the website describes Indigenous scientists and engineers in Canada and the United States.
http://dream-catching.com. and http://www.nativeaccess.com. Retrieved Sept. 14, 2010.

Role models: http://www.nativeaccess.com/allabout/rolemodels.html. Retrieved Sept. 14, 2010.

Government of Nunavut
The Department of Human Resources, Government of Nunavut (Canada) has produced a website to describe the essence of traditional Inuit knowledge—Inuit Qaujimajatuqangit (IQ). The goal is to integrate IQ into all government programs and services, including school science.
http://www.gov.nu.ca/hr/site/beliefsystem.htm. Retrieved Sept. 14, 2010.

Rekindling Traditions
Six cross-cultural science and technology units, a teachers' guide, and other resources were developed through collaboration among teachers, the University of Saskatchewan, and community Knowledge Keepers and Indigenous families across northern Saskatchewan (see Appendix A).
http://www.usask.ca/education/ccstu/. Retrieved Sept. 14, 2010.

Resource Management
The *Forests and Oceans for the Future* project helps incorporate Indigenous and non-Indigenous core community values and knowledge in local sustainable forest and natural resource management. It designs educational materials to facilitate mutual respect and knowledge sharing between First Nations and others.
http://www.ecoknow.ca/. Retrieved Sept. 14, 2010.

USA

Alaskan Native Science Commission
General information on traditional knowledge systems in the Artic is available here. These systems are also compared to scientific knowledge systems.
http://www.nativescience.org/issues/tk.htm. Retrieved Sept. 14, 2010.

Alaskan Native Knowledge Network
The ANKN serves as a major resource for compiling and exchanging information related to Alaska Native knowledge systems and ways of knowing in school science. Publications include *Culturally Responsive Schools*, *Preparing Culturally Responsive Teachers for Alaska's Schools*, *Handbook for Culturally Responsive Science Curriculum*, and *Guidelines for Cross-Cultural Orientation Programs*.
http://www.ankn.uaf.edu/. This and the site below were retrieved Sept. 14, 2010.

Besides these documents, CDs and other resources may be helpful, such as http://www.ankn.uaf.edu/sop/. Bulletin *Sharing Our Pathways*.
http://www.ankn.uaf.edu/Curriculum/Units/. Science teaching units (about 21 of them).
http://ankn.uaf.edu/curriculum/Articles/RayBarnhardt/TLAC.html. The excellent article "Teaching/Learning across Cultures: Strategies for Success."

Ancient Observations, Timeless Knowledge
This NASA website disseminates information on curriculum projects related to Native Americans, and biographies of Native Americans working at NASA. http://sunearthday.nasa.gov/2005/na/index.htm. Retrieved Sept. 14, 2010.

The Mystery of Chaco Canyon
The Solstice Project is dedicated to the study of ancient cultures of the American Southwest. The Project was founded in 1978 to study, document, and preserve the remarkable Sun Dagger—a celestial calendar of the ancient Pueblo people—and other notable achievements of ancient Southwestern culture. This celestial calendar technology, with no moving parts, simultaneously keeps track of the movement of the Sun and Moon. No technology before the computer has accomplished this feat. http://www.solsticeproject.org/. Retrieved Sept. 14, 2010.

The Cradleboard Teaching Project
This project informs public education about Native American culture. It was founded by Buffy Sainte-Marie in 1996. The CD-ROM *Science through Native American Eyes* does not work on all computers, unfortunately. Inquire before purchasing.
http://www.cradleboard.org/. Retrieved Sept. 14, 2010.

Indigenous Science Resources: Digital Library
This website links to many resources, mostly in the United States. http://www.dlisr.org/index.html. Retrieved Sept. 14, 2010.

AUSTRALIA

Australian Academy of Science
PrimaryConnections: Linking Science with Literacy
The "Indigenous Perspectives" section of PrimaryConnections includes
- the Indigenous Perspectives Framework with information and resource links,
- Indigenous Perspectives curriculum links for a suite of *PrimaryConnections* units,
- a professional learning module,

- links to each chapter of the *Connecting Minds* DVD, and
- the Indigenous Perspectives pilot study research report: *Small Study—Big Success Story.*

http://www.science.org.au/primaryconnections/indigenous. Retrieved Sept. 14, 2010.

Living Knowledge: **Indigenous Knowledge** *in Science Education*
This website indicates the most effective ways of incorporating Indigenous knowledge within the secondary school science curricula in the state of New South Wales, Australia.
http://livingknowledge.anu.edu.au/. Retrieved Sept. 14, 2010.

GLOBAL

Indigenous Science Network
This network produces a bimonthly *ISN Bulletins* exchange of information and ideas, archived at the website.
http://members.ozemail.com.au/~mmichie/network.html. Retrieved Sept. 14, 2010.

UNESCO
The purpose of the United Nations Educational, Scientific, and Cultural Organization (UNESCO) publication *Education for Sustainable Development* is to integrate the principles, values, and practices of sustainable development into all aspects of education and learning to encourage changes in behaviour that will create a more sustainable future in terms of environmental integrity, economic viability, and a just society for present and future generations.
http://www.unesco.org/en/esd/. Retrieved Sept. 14, 2010.

APPENDIX D
Recommended Books about Indigenous Worldviews

The academic references that reappear in this book are useful for looking into a specific issue of interest. In addition to any of these sources, we highly recommend the following books that present an Indigenous perspective in a more personal way, in greater detail, and from a different perspective than this book has. Consider Appendix D as the start to a growing list of excellent reads that colleagues may suggest during your conversations about Indigenous ways of living and being.

Greg Cajete; ***Look to the Mountain: An Ecology of Indigenous Education***, 1994; Kivaki Press (Skyland, NC).
"Look to the mountain" is a metaphor in Indigenous-based education. This is a 'must read' for teachers who seek to deepen their understanding on the ecological nature of Indigenous thought and culture. Cajete focuses on place-based knowledge and he organizes it around a set of concentric circles that define a framework for Indigenous education.

Greg Cajete; ***Native Science: Natural Laws of Interdependence***, 2000; Clear Light Publishers (Sante Fe, NM).
This groundbreaking reader is important for all cross-cultural science teachers. Cajete's ideas are applicable to any context that seeks to foster an ecological consciousness in school science. Holistically, he deals with art, myth, ceremony, and symbolism integrated into an Indigenous understanding of reality—Native science.

Ward Churchill; ***Struggle for the Land***, 1999; Arbeiter Ring (Winnipeg, MB).
This book was awarded the Gustavus Myers Award for literature on human rights. From an Indigenous standpoint, Churchill documents historical events on Turtle Island (North America) since the time Europeans arrived. Although colonization is chronicled in terms of theft, deception, and genocide, the book primarily speaks to the resistance, strength, and resilience of Indigenous peoples and their worldviews.

Wade Davis; *The Wayfinders: Why Ancient Wisdom Matters in the Modern World*, 2009; Anansi (Toronto, ON).
Davis details his personal in-depth contact with the worldviews held by descendants of

- the first humans to inhabit planet Earth (the San of Africa),
- the first wave of humans to leave Africa (the Indigenous peoples of Australia),
- the Polynesians who first voyaged among the Pacific islands,
- the Inka Empire (the Quechua peoples of the Andes in South America),
- the Tairona of Columbia's Sierra Nevada de Santa Marta, and
- First Nations peoples of British Columbia, Canada.

Davis explains a current controversy taking place between industrial progress and Indigenous land rights, in terms of Indigenous worldviews that clash with Eurocentric ones.

Vine Deloria; *Red Earth, White Lies: Native Americans and the Myth of the Scientific Fact*, 1995; Scribner (New York, NY).
Deloria's book is an excellent source for understanding a thoughtful Indigenous account of certain politics and myths found in Eurocentric science. This leading Native American scholar contrasts Eurocentric theory about the world with the ancestral worldviews of Native Americans. Based on Eurocentric science and Indigenous evidence, he challenges scientific theories about prehistoric North America, such as the Bering Strait Theory, and offers alternative views.

Terry Garvin; *Carving Faces, Carving Lives: People of the Boreal Forest*, 2005; Heritage Community Foundation (Edmonton, AB).
Through extraordinary colour photographs and detailed information, Garvin describes knowledge generally held by Indigenous peoples inhabiting the Boreal Forest across Canada. Topics include camp and home lifestyle, forests, wildlife, travel, and art and crafts. A Cree/English glossary is included. It is available directly from the author: garvints@shaw.ca or (403-286-1334).

Thomas King; *Green Grass, Running Water*, 1994; Harper Perennial (Toronto, ON).
Award-winning novelist and creator of the CBC radio series *Dead Dog Cafe*, Cherokee Thomas King blends infectious humour with Indigenous

philosophy by following an extended family to a Blackfoot Sun Dance. The encounters of Eli, Alberta, Lionel, as well as other characters, particularly four very old Elders and their companion Coyote (the trickster), provide insight into King's Indigenous worldview. "Stay calm. Be brave. Wait for the signs."

Kathleen D. Moore, Kurt Peters, Ted Jojola, and Amber Lacey; *How It Is: The Native American Philosophy of V.F. Cordova*, 2007; University of Arizona Press (Tucson, AZ).
This posthumous collection of unpublished papers, poems, and a short story by Viola Faye Cordova (1937–2002) provides an understanding of Indigenous worldviews, the nature of reality, the origins of the world, the relation between matter and spirit, and the nature of time. These topics are tied together by one theme—the role of culture and language in a person's life.

Ed McGaa (Eagle Man); *Mother Earth Spirituality: Native American Paths to Healing Ourselves and Our World*, 1990; Harper Collins Publishers (New York, NY).
Writing from an Oglala perspective, Eagle Man illustrates the connections among Indigenous philosophies, rituals, and the importance of reconnecting with the earth in a spiritual way. He explains different types of ceremonies in Indigenous cultures.

Melissa Nelson; *Original Instructions: Indigenous Teachings for a Sustainable Future*, 2008; Bear & Company (Rochester, VT).
This excellent book of Indigenous teachings has a strong focus on sustainable ways of living and being. The author showcases perspectives of more than 30 contemporary Indigenous leaders.

Carol Schaefer; *Grandmothers Counsel the World: Women Elders Offer Their Vision for Our Planet*, 2006; Trumpeter Books (Boston, MA).
This is a collection of stories and 'earth wisdom' shared by a group of diverse Indigenous women Elders. Schaefer highlights the importance of Indigenous women's ways of knowing with respect to environmental issues and sustainable ways of living. This will be most helpful to teachers looking for potent quotes concerning Indigenous teachings.

Richard Wagamese; *One Native Life*; 2008; Douglas & McIntyre (Vancouver, BC).

Each short chapter eloquently describes a positive event that helped Wagamese during his journey to become aware of, and then reclaim, his Ojibwa culture of northwestern Ontario. Abandoned as a child, a ward of foster care, and then adopted by a white family, he grew up during painstaking years and became an established novelist. From the beauty of his rustic lakeside home in the interior of British Columbia, he recalls and reflects on events in his life, revealing aspects of his Indigenous worldview and lessons for school teachers about how to help Indigenous students strengthen their cultural identities.

REFERENCES

Aikenhead, G.S. (2000). *Rekindling traditions: Cross-cultural science & technology units.* Retrieved September 14, 2010, from http://www.usask.ca/education/ccstu/.

Aikenhead, G.S. (2002). Cross-cultural science teaching: Rekindling traditions for Aboriginal students. *Canadian Journal of Science, Mathematics and Technology Education, 2,* 287–304.

Aikenhead, G.S. (2005). Science-based occupations and the science curriculum: Concepts of evidence. *Science Education, 89,* 242–275.

Aikenhead, G.S. (2006). *Science education for everyday life: Evidence-based practice.* New York, NY: Teachers College Press.

Aikenhead, G.S., & Elliott, D. (2010). An emerging decolonizing science education in Canada. *Canadian Journal of Science, Mathematics and Technology Education, 11,* in press.

Aikenhead, G.S., & Jegede, O.J. (1999). Cross-cultural science education: A cognitive explanation of a cultural phenomenon. *Journal of Research in Science Teaching, 36,* 269–287.

Aikenhead, G.S., & Ogawa, M. (2007). Indigenous knowledge and science revisited. *Cultural Studies of Science Education, 2,* 539–591.

Akatugba, A.H., & Wallace, J. (2009). An integrative perspective on students' proportional reasoning in high school physics in a West African context. *International Journal of Science Education, 31,* 1473–1493.

Albe, V. (2008). Students' positions and considerations of scientific evidence about a controversial socioscientific issue. *Science & Education, 17,* 805–827.

Alaska Native Science Commission. (2009). *What is traditional knowledge?* Retrieved September 14, 2010, from http://www.nativescience.org/html/traditional_knowledge.html.

American Association for the Advancement of Science (AAAS). (1977). *Native Americans in science.* Washington, DC: AAAS.

ANKN (Alaska Native Knowledge Network). (1996). *Spiral pathway for integrating rural Alaska learning.* Retrieved September 14, 2010, from http://www.ankn.uaf.edu/.

Atleo, E.R. (2004). *Tsawalk: A Nuu-chah-nulth worldview.* Vancouver, BC: UBC Press.

Australian Academy of Science. (2009). *Primary Connections: Indigenous perspectives: Teaching and learning guide.* Retrieved September 14, 2010, from http://www.science.org.au/primaryconnections/ip-qtl.htm.

Barnhardt, R. (2006). *Teaching/learning across cultures: Strategies for success.* Retrieved September 14, 2010, from http://ankn.uaf.edu/curriculum/Articles/RayBarnhardt/TLAC.html.

Barnhardt, R., & Kawagley, A.O. (2005). Indigenous knowledge systems and Alaska Native ways of knowing. *Anthropology and Education Quarterly, 36,* 8–23.

Barnhardt, R., Kawagley, A.O., & Hill, F. (2000). Cultural standards and test scores. *Sharing Our Pathways, 5*(4), 1–4.

Bastein, G. (2004). *Blackfoot ways of knowing.* Calgary, AB: University of Calgary Press.

Battiste, M. (Ed.). (2000). *Reclaiming Indigenous voice and vision.* Vancouver, BC: University of British Columbia Press.

Battiste, M. (2002). *Indigenous knowledge and pedagogy in First Nations education: A literature review with recommendations*. Ottawa, ON: Indian and Northern Affairs Canada.

Battiste, M., & Henderson, J.Y. (2000). *Protecting Indigenous knowledge and heritage*. Saskatoon, SK: Purich Publishing.

Bauer, H.H. (1992). *Scientific literacy and the myth of the scientific method*. Chicago, IL: University of Illinois Press.

Belczewski, A. (2009). Decolonizing science education and the science teacher: A White teacher's perspective. *Canadian Journal of Science, Mathematics and Technology Education, 9*, 191–202.

Bell, B., & Cowie, B. (2001). *Formative assessment and science education*. Dordrecht, The Netherlands: Kluwer Academic Publishers.

Bencze, J.L. (2008). Private profit, science, and science education: Critical problems and possibilities for action. *Canadian Journal of Science, Mathematics, and Technology Education, 8*, 297–312.

Berger, P.L., & Luckmann, T. (1966). *The social construction of reality*. New York, NY: Anchor Books.

Bird, L. (2010). *Omushkego oral history project*. Retrieved July 20, 2010, from http://www.ourvoices.ca.

Blackwater, A. (2009, May). *Personal perspectives of Aboriginal science*. A presentation to the Aboriginal Science Symposium, University of Lethbridge, Alberta.

Bolter, J.D. (1984). *Turing's man: Western culture in the computer age*. New York, NY: Viking Penguin Inc.

Boulton, J., Brockman, A., Johanson, T., Wallace, M., & View, T. (2010). *Pearson Saskatchewan science 8*. Toronto, ON: Pearson Canada Inc.

Bowden, C. (2010). Native lands. *National Geographic, 218* (August), 80–97.

Brandt, C. (2008). Scientific discourse in the academy. A case study of an American Indian undergraduate. *Science Education, 92*, 825–847.

Brayboy, B.M.J., & Castagno, A.E. (2008). How might Native science inform "informal science learning"? *Cultural Studies of Science Education, 3*, 731–750.

Brockman, A., Doepker, C., Stephenson, E., Wallace, M., & View, T. (2009). *Pearson Saskatchewan science 7*. Toronto, ON: Pearson Canada Inc.

Brotman, J.S., & Moore, F.M. (2008). Girls and science: A review of four themes in the science education literature. *Journal of Research in Science Teaching, 45*, 971–1002.

Bull, R. (2008). *Small study—big success story* (Report by the Australian Academy of Science). Retrieved August 24, 2010, from http://www.science.org.au/primaryconnections/indigenous/images/ip-report.pdf.

Cajete, G. (1994). *Look to the Mountain: An ecology of Indigenous education*. Durango, CO: Kivaki Press.

Cajete, G.A. (1999). *Igniting the sparkle: An Indigenous science education model*. Skyand, NC: Kivaki Press.

Cajete, G. (2000a). Indigenous knowledge: The Pueblo metaphor of Indigenous education. In M. Battiste (Ed.), *Reclaiming Indigenous voice and vision* (pp. 181–191). Vancouver, BC: University of British Columbia Press.

Cajete, G. (2000b). *Native science: Natural laws of interdependence*. Santa Fe, NM: Clear Light.

Cajete, G. (2006). Western science and the loss of natural creativity. In Four Arrows (a.k.a. D.T. Jacobs) (Ed.), *Unlearning the language of conquest: Scholars expose anti-Indianism in America* (pp. 247–259). Austin, TX: University of Texas Press.

Cajori, F. (Translator) (1962). *Sir Isaac Newton's mathematical principles of natural philosophy and his system of the world (Principia)*. Berkeley, CA: University of California Press.

Canadian Council on Learning. (2007a). *Redefining how success is measured in First Nations, Inuit and Métis learning*. Retrieved April 23, 2009, from http://www.ccl-cca.ca/CCL/Reports/RedefiningSuccessInAboriginalLearning/?Language=EN.

Canadian Council on Learning. (2007b). *The cultural divide in science education for aboriginal learners*. Retrieved April 23, 2009, from http://www.ccl-cca.ca/CCL/Reports/LessonsInLearning/LinL20070116_Ab_sci_edu.htm.

Capra, F. (1996). *The web of life: A new scientific understanding of living systems*. New York, NY: Doubleday.

Castellano, M.B. (2000). Updating Aboriginal traditions of knowledge. In G.J.S. Dei, B.L. Hall, & D.G. Rosenberg (Eds.), *Indigenous knowledges in global contexts: Multiple readings of our world* (pp. 1–36). Toronto, ON: University of Toronto Press.

CBC (Canadian Broadcasting Corporation). (1995). *Thunderbirds*. Toronto, ON: Author.

CBC (Canadian Broadcasting Corporation). (2003). *Mother Earth*. Toronto, ON: Author.

CBU (Cape Breton University). (2007). *Two-eyed seeing* (video). Sydney, NS: Institute for Integrative Science & Health. Retrieved September 1, 2010, from http://video.google.com/videoplay?docid=-4641523553883632087.

Chalmers, A.F. (1999). *What is this thing called science?* (3rd ed.). Birmingham, UK: Open University Press.

Chang, P.J., & Rosiek, J. (2003). Anti-colonialist antinomies in a biology lesson: A sonata-form case study of cultural conflict in a science classroom. *Curriculum Inquiry, 33*, 251–290.

Chinn, P.W.U. (2007). Decolonizing methodologies and Indigenous knowledge: The role of culture, place and personal experience in professional development. *Journal of Research in Science Teaching, 44*, 1247–1268.

Chinn, P.W.U. (2008). *Malama I Ka 'Aina: Sustainability through Traditional Hawaiian Practices*. Honolulu, HI: University of Hawai'i at Manoa. Retrieved June 9, 2009, from http://malama.hawaii.edu/.

Clark, W.C., & Dickson, N.M. (2003). Sustainability science: The emerging research program. *Proceedings of the National Academy of Sciences of the United States of America, 100*(14), 8049–8061.

Coalition for the Advancement of Aboriginal Studies. (2002). *Learning about walking in beauty: Placing Aboriginal perspectives in Canadian classrooms*. Toronto, ON: Canadian Race Relations Foundation.

Cobern, W.W. (1996). Worldview theory and conceptual change in science education. *Science Education, 80*, 579–610.

Cole, S. (1992). *Making science: Between nature and society*. Cambridge, MA: Harvard University Press.

Cobern, W.W. (2000). *Everyday thoughts about nature*. Boston, MA: Kluwer Academic.

Collingridge, D. (1989). Incremental decision making in technological innovations: What role for science? *Science, Technology, & Human Values, 14*, 141–162.

Constantinou, C., Hadjilouca, R., & Papadouris, N. (2010). Students' epistemological awareness concerning the distinction between science and technology. *International Journal of Science Education, 32*, 143–172.

Council of Ministers of Education Canada. (1997). *Common framework of science learning outcomes: Pan-Canadian protocol for collaboration on school curriculum*. Ottawa, ON: Author.

Council of Ministers of Education Canada. (2002). *Best practices in increasing Aboriginal postsecondary enrolment rates*. Victoria, BC: R.A. Malatest & Associates Ltd.

CRYSTAL (Centres for Research in Youth, Science Training and Learning). (2010). Nunavut resources. Ottawa: Natural Sciences and Engineering Research Council of Canada (NSERC). Retrieved January 25, 2011, from http://www.umanitoba.ca/outreach/crystal/nunavut.html.

Cuthand, D. (2007). *Askiwina: A Cree world*. Regina, SK: Coteau Books.

Davis, W. (2009). *The wayfinders: Why ancient wisdom matters in the modern world*. Toronto, ON: Anansi.

Deloria, V. (1992). Relativity, relatedness and reality. *Winds of Change, 7* (Autumn), 35–40.

DeMerchant, R.V. (2002). *A case study of integrating Inuuqatigiit into a Nunavut junior high school classroom*. Unpublished master's thesis, University of Saskatchewan, Saskatoon, SK.

Department of Human Resources (Government of Nunavut). (2005). *Inuit Qaujimajatuqangit*. Retrieved September 14, 2010, from http://www.gov.nu.ca/hr/site/beliefsystem.htm.

Devall, B., & Sessions, G. (1999). Deep ecology. In M.J. Smith (Ed.), *Thinking through the environment: A reader* (pp. 200–208). New York, NY: Routledge.

Djerassi, C. (1998). Ethical discourse by science-in-fiction. *Nature, 393*, 11 June, 511.

Duran, P.H. (2007). On the cosmic order of modern physics and the conceptual world of the American Indian. *World Futures, 63*, 1–27.

Dyck, L. (1998). An analysis of Western, feminist and Aboriginal science using the medicine wheel of the Plains Indians. In L.A. Stiffarm (Ed.), *As we see ...: Aboriginal pedagogy* (pp. 87–101). Saskatoon, SK: University Extension Press, University of Saskatchewan.

Einstein, A. (1956). *Albert Einstein: Out of my later years*. New York, NY: Winds Books.

Eliot, T.S. (1963). *Choruses from the rock: Collected poems, 1909–1962*. London, UK: Faber.

Elliott, F. (2008). *Western science coming-to-know traditional knowledge*. An unpublished doctoral dissertation. University of Alberta, Edmonton, AB.

Environics Institute. (2010). *The urban Aboriginal peoples study: Background and summary of main findings*. Retrieved April 24, 2010, from http://www.uaps.ca/wp-content/uploads/2010/04/UAPS-Report-Summary.pdf.

Ermine, W.J. (1995). Aboriginal epistemology. In M. Battiste & J. Barman (Eds.), *First Nations education in Canada: The circle unfolds* (pp. 101–112). Vancouver, BC: University of British Columbia Press.

Ermine, W.J. (1998). Pedagogy from the ethos: An interview with Elder Ermine on language. In L.A. Stiffarm (Ed.), *As we see ... Aboriginal pedagogy* (pp. 9–28). Saskatoon, SK: University Extension Press, University of Saskatchewan.

Four Arrows (aka D.T. Jacobs) (Ed.) (2006). American Indian worldviews and values. In Four Arrows (D.T. Jacobs) (Ed.), *Unlearning the language of conquest: Scholars expose anti-Indianism in America* (pp. 278–280). Austin, TX: University of Texas Press.

Fourez, G. (1989). Scientific literacy, societal choices, and ideologies. In A.B. Champagne, B.E. Lovitts, & B.J. Calinger (Eds.), *Scientific literacy* (pp. 89–108). Washington, DC: American Association for the Advancement of Science.

FreshWater Summit. (2010). F. Henry Lickers. Retrieved August 20, 2010, from http://2010freshwatersummit.org/speakers.htm.

FSIN (Federation of Saskatchewan Indian Nations). (2009). "FSIN science fair judge's form A." Saskatoon, Saskatchewan: FSIN. Retrieved February 15, 2010, from http://www.fsin.com/index.php/science-program.html.

Fulbright, J.W. (1964). *Old myths and new realities.* New York, NY: Random House.

Galeano, E. (1973). *Open veins of Latin America: Five centuries of the pillage of a continent.* New York, NY: Monthly Review Press.

Garvin, T. (2005). *Carving faces, carving lives: People of the Boreal Forest.* Edmonton, AB: Heritage Community Foundation.

Gaskell, P.J. (1992). Authentic science and school science. *International Journal of Science Education, 14,* 265–272.

Gaskell, P.J. (2003). Engaging science education within diverse cultures. *Curriculum Inquiry, 33,* 235–249.

George, J.M. (1999). Worldview analysis of knowledge in a rural village: Implications for science education. *Science Education, 83,* 77–95.

Gieryn, T.F. (1999). *Cultural boundaries of science: Credibility on the line.* Chicago, IL: University of Chicago Press.

Glasson, G.E., Frykholm, J.A., Mhango, B.A, & Phiri, A.D. (2006). Understanding the earth systems of Malawi: Ecological sustainability, culture, and place-based education. *Science Education, 90,* 660–680.

Gleick, P.H., et al. (2010). Climate change and the integrity of science. *Science, 328* (May 7), 689–670. Retrieved May 14, 2010, from www.sciencemag.org.

Goldstein, M., & Goldstein, I.F. (1981). *How we know: An exploration of the scientific process.* New York, NY: Da Capo Press.

Gott, R., Duggan, S., & Roberts, R. (2007). *Concepts of evidence.* University of Durham, UK. Retrieved August 20, 2010, from http://www.dur.ac.uk/rosalyn.roberts/Evidence/CofEv_Gott%20et%20al.pdf.

Gott, R., Duggan, S., Roberts, R., & Hussain, A. (2009). *Research into understanding scientific evidence.* University of Durham, UK. Retrieved August 20, 2010, from http://www.dur.ac.uk/richard.gott/Evidence/cofev.htm.

Gough, N. (2002). Thinking/acting locally/globally: Western science and environmental education in a global knowledge economy. *International Journal of Science Education, 24,* 1217–1237.

Greene, B. (1999). *The elegant universe.* New York, NY: W.W. Norton & Company.

Habermas, J. (1972). *Knowledge and human interests.* London, UK: Heinemman.

Hampton, E. (1995). Towards a redefinition of Indian education. In M. Battiste & J. Barman (Eds.), *First Nations education in Canada: The circle unfolds* (pp. 5–46). Vancouver, BC: University of British Columbia Press.

Hampton, M., & Roy, J. (2002). Strategies for facilitating success of First Nations students. *Canadian Journal of Higher Education, 32*(3), 1–28.

Harding, S. (1998). Multiculturalism, postcolonialism, feminism: Do they require new research epistemologies? *Australian Educational Research, 25*(1), 37–51.

Hatcher, A., Bartlett, C., Marshall, A., & Marshall, M. (2009a). Two-Eyed Seeing: A cross-cultural science journey. *Green Teacher* (86), 3–6.

Hatcher, A., Bartlett, C., Marshall, A., & Marshall, M. (2009b). Two-Eyed Seeing in the classroom environment: Concepts, approaches, and challenges. *Canadian Journal of Science, Mathematics and Technology Education, 9,* 141–153.

Hazen, R.M. (2005). *Genesis: The scientific quest for life's origin.* Washington, DC: Joseph Henry Press.

Herbert, S. (2008). Collateral learning in science: Students' responses to a cross-cultural unit of work. *International Journal of Science Education, 30,* 979–993.

Hodson, D. (1998). Science fiction: The continuing misrepresentation of science in the school curriculum. *Curriculum Studies, 6*, 191–215.

Hodson, D. (2009). *Teaching and learning about science: Language, theories, methods, history, traditions and values.* Boston, MA: Sense Publishers.

Holton, G. (1978). *The scientific imagination: Case studies.* Cambridge, UK: Cambridge University Press.

Hounjet, C., Kvamme, B., Mohr, P., Phillipchuk, K., & View, T. (2011). *Pearson Saskatchewan science 9.* Toronto, ON: Pearson Canada Inc.

Hughes, P., More, A.J., & Williams, M. (2004). *Aboriginal ways of learning.* Adelaide, Australia: Flinders Press.

Hunt, B., & Harrington, C.F. (2008). The impending educational crisis for American Indians: Higher education at the crossroads. *Journal of Multicultural, Gender and Minority Studies, 2*(2), 1–11.

Hutchison, P., & Hammer, D. (2010). Attending to student epistemological framing in a science classroom. *Science Education, 94*, 506–524.

Ignas, V. (2004). Opening doors to the future: Applying local knowledge in curriculum development. *Canadian Journal of Native Education, 28*, 49–60.

Indigenous Education Institute. (2009). Current highlights. Retrieved September 14, 2010, from http://www.indigenouseducation.org/ index.html.

International Council for Science. (2002). *Science, traditional knowledge and sustainable Development.* Paris, France: Author.

Inuit Subject Advisory Committee. (1996). *Inuuqatigiit: The curriculum from the Inuit perspective.* Yellowknife, NWT: Department of Education, Culture and Employment.

Ipellie, A. (2007). *The Inuit thought of it: Amazing arctic innovations.* Toronto, ON: Annick Press.

Irzik, G. (1998). Philosophy of science and radical intellectual Islam in Turkey. In W.W. Cobern (Ed.), *Socio-cultural perspectives on science education* (pp. 163–179). Boston, MA: Kluwer Academic.

James, K. (2001). Fires need fuel: Merging science education with American Indian community needs. In K. James (Ed.), *Science and Native American communities: Legacies of pain, visions of promise* (pp. 2–8). Lincoln, NE: University of Nebraska Press.

Jegede, O.J., & Okebukola, P.A. (1991). The relationship between African traditional cosmology and students' acquisition of a science process skill. *International Journal of Science Education, 13*, 37–47.

Johanson, T, Mohr, P., Treptau, C., Wallace, C., & View, T. (2009). *Pearson Saskatchewan science 6.* Toronto, ON: Pearson Canada Inc.

Kanu, Y. (2002). In their own voices: First Nations students identify some cultural mediators of their learning in the formal school system. *Alberta Journal of Educational Research, 48*, 98–121.

Kawagley, A.O. (1990). Yup'ik ways of knowing. *Canadian Journal of Native Education, 17*(2), 5–17.

Kawagley, A.O. (1995). *A Yupiaq worldview.* Prospect Heights, IL: Waveland Press.

Kawagley, A.O., Norris-Tull, D., & Norris-Tull, R.A. (1998). The indigenous worldview of Yupiaq culture: Its scientific nature and relevance to the practice and teaching of science. *Journal of Research in Science Teaching, 35*, 133–144.

Kawasaki, K. (2002). A cross-cultural comparison of English and Japanese linguistic assumptions influencing pupils' learning of science. *Canadian and International Education, 31*, 19–51.

Keane, M. (2008). Science education and worldview. *Cultural Studies of Science Education, 3*, 587–613.

Kelly, G.J., Carlsen, W.S., & Cunningham, C.M. (1993). Science education in sociocultural context: Perspectives from the sociology of science. *Science Education, 77*, 207–220.

Knight, D. (2001). *The seven fires: Teachings of the Bear Clan.* Muskoday First Nation, Saskatchewan, SK: Many Worlds Publishing.

Knudtson, P., & Suzuki, D. (1992). *Wisdom of the elders.* Toronto, ON: Stoddart.

Kuhn, T. (1962/1970). *The structure of scientific revolutions* (2nd ed. in 1970). Chicago, IL: University of Chicago Press.

Landon, R. (2008). *A Native American thought of it: Amazing inventions and innovations.* Toronto, ON: Annick Press.

Lee, O. (2002). Promoting scientific inquiry with elementary students from diverse cultures and languages. *Review of Research in Education, 26*, 23–69.

Levinson, R. (2010). Science education and democratic participations: An uneasy congruence? *Studies in Science Education, 46*, 69–119.

Lewontin, R.C. (1991). *Biology as ideology: The doctrine of DNA.* Concord, ON: House of Anansi Press.

Lewthwaite, G., & McMillan, B. (2007). Combing the views of both worlds: Perceived constraints and contributors to achieving aspirations for science education in Qikiqtani. *Canadian Journal of Science, Mathematics and Technology Education, 7*, 355–376.

Lewthwaite, G., & McMillan, B. (2010). "She can bother me, and that's because she cares:" What Inuit students say about teaching and their learning. *Canadian Journal of Education, 33*, 140–175.

Little Bear, L. (2000). Jagged worldviews colliding. In M. Battiste (Ed.), *Reclaiming Indigenous voice and vision* (pp. 77–85). Vancouver, BC: University of British Columbia Press.

Little Bear, L. (2009, May). *The hidden science.* A presentation to the Aboriginal Science Symposium, University of Lethbridge, Alberta.

Longino, H. (1990). *Science as social knowledge: Values and objectivity in scientific inquiry.* Princeton, NJ: Princeton University Press.

Loo, S.P. (2001). Islam, science and science education: Conflict or concord? *Studies in Science Education, 36*, 45–78.

Loo, S.P. (2007). The two cultures of science: On language-culture incommensurability concerning 'nature' and 'observation.' *Higher Education Policy, 20*, 97–116.

Lubben, F., & Campbell, B. (1996). Contextualizing science teaching in Swaziland: Some student reactions. *International Journal of Science Education, 18*, 311–320.

Lyver, P.O'B., Jones, C., & Moller, H. (2009). Looking past the wallpaper: Considerate evaluation of traditional environmental knowledge by science. *Journal of the Royal Society of New Zealand, 39*, 219–223.

MacIvor, M. (1995). Redefining science education for Aboriginal students. In M. Battiste & J. Barman (Eds.), *First Nations education in Canada: The circle unfolds* (pp. 73–98). Vancouver, BC: University of British Columbia Press.

MacLeod, R., & Collins, P. (Eds.) (1981). *The parliament of science.* Northwood, UK: Science Reviews.

Manitoba Education and Youth (2003). *Integrating Aboriginal perspectives into Curricula.* Winnipeg, Manitoba: Author. Retrieved February 19, 2010 from http://www.edu.gov.mb.ca/k12/docs/ policy/abpersp/ab_persp.pdf.

Mäori Proverbs (no author). (1992). *Mäori Proverbs.* Birkenhead, Auckland, NZ: Reed Publishing.

Marshall, A. (2007, May). *Two-Eyed Seeing.* A presentation to the Two-Eyed Seeing Workshop for Science Education for Children and Youth. Cape Breton University, Nova Scotia, Canada. Retrieved September 14, 2010, from http://marcatodigital.com/iish/pdf/may_24_workshop_summary.pdf.

Maryboy, N.C., & Begay, D.H. (1998). *Living the order: Dynamic cosmic process of Diné cosmology. Nanit'a saah naaghai nanit'a bik'eh hozhoon.* Unpublished Ph.D. dissertation, California Institute of Integral Studies, San Francisco, CA. ProQuest website, retrieved January 24, 2011, from http://proquest.umi.com/pqdlink?Ver=1&Exp=01-24-2016&FMT=7&DID=733466661&RQT=309&attempt=1&cfc=1.

Maryboy, N.C., & Begay, D.H. (2005). *Sharing the skies: Navajo astronomy—a cross-cultural view.* Bluff, UT: Indigenous Education Institute.

Maryboy, N.C., Begay, D.H., & Nichol, L. (2006). Paradox and transformation. *World Indigenous Nations Higher Education Consortium, Volume 2.* Retrieved May 25, 2010, from http://www.indigenouseducation.org/WINHEC%20Journal%203-29-06%20Final%20c.pdf.

Matthews, C., & Smith, W. (1994). Native American related materials in elementary science instruction. *Journal of Research in Science Teaching, 41,* 363–380.

McClune, B., & Jarman, R. (2010). Critical reading of science-based news reports: Establishing a knowledge, skills and attitudes framework. *International Journal of Science Education, 32,* 727–752.

McGilchrist, I. (2009). *The master and his emissary: The divided brain and the making of the western world.* New Haven, CT: Yale University Press.

McGregor, D. (2000). The state of traditional ecological knowledge research in Canada: A critique of current theory and practice. In R.F. Laliberte, P. Settee, J.B. Waldram, R. Innes, B. Macdougall, L. McBain & F.L. Barron (Eds.), *Expressions in Canadian Native Studies* (pp. 436–458). Saskatoon, SK: University of Saskatchewan Extension Press.

McGregor, D. (2002). Traditional ecological knowledge and the two-row wampum. *Biodiversity, 3*(3), 2–3.

McKinley, E. (1996). Towards an indigenous science curriculum. *Research in Science Education, 26,* 155–167.

McKinley, E. (2005). Locating the global: Culture, language and science education for indigenous students. *International Journal of Science Education, 27,* 227–241.

McKinley, E. (2007). Postcolonialism, indigenous students, and science education. In S.K. Abell & N.G. Lederman (Eds.), *Handbook of research on science education* (pp. 199–226). Mahwah, NJ: Lawrence Erlbaum.

McKinley, E., Stewart, G., & Richards, P. (2004). *Māori knowledge, language and participation in mathematics and science education.* (Final Report). Hamilton, Aotearoa New Zealand: University of Waikato, School of Education.

McNeill, K.L. (2009). Teachers' use of curriculum to support students in writing scientific arguments to explain phenomena. *Science Education, 93,* 233–268.

Mendelsohn, E. (1976). Values and science: A critical reassessment. *The Science Teacher, 43*(1), 20–23.

Menzies, C.R. (Project Leader) (2003). *Forests for the future.* Vancouver, BC: Department of Anthropology, University of British Columbia. Retrieved February 27, 2010, from http://www.ecoknow.ca.

Menzies, C. (2006). *Traditional ecological knowledge and natural resource management.* Lincoln, NE: Univ of Nebraska Press.

Michell, H. (2005). Nîhîthêwâk of Reindeer Lake, Canada: Worldview, epistemology, and relationships with the natural world. *Australian Journal of Indigenous Education, 43,* 33–43.

Michell, H. (2007). *Nîhîthewâk Ithîníwak and Science Education: An Exploratory Narrative Study Examining Indigenous-based Science Education in K–12 Classrooms from the Perspectives of Teachers in Woodlands Cree Community Contexts.* Unpublished Ph.D. dissertation, University of Regina, Regina, SK. ProQuest website, retrieved November 24, 2010, from http://gradworks.umi.com/NR/42/NR42507.html.

Michell, H., Vizina, Y., Augustus, C., & Sawyer, J. (2008). *Learning Indigenous science from place: Research study examining Indigenous-based science perspectives in Saskatchewan First Nations and Métis community contexts.* Ottawa, ON: Canadian Council on Learning. Retrieved March 10, 2010, from http://www.ccl-cca.ca/pdfs/FundedResearch/Michell-FinalReport-14Nov2008-AbL2006.pdf.

Michie, M. (2002). Why Indigenous science should be included in the school science curriculum. *Australian Science Teachers' Journal, 48*(2), 36–40.

Milne, C.E., & Taylor, P.C. (1998). Between myth and a hard place. In W.W. Cobern (Ed.), *Socio-cultural perspectives on science education* (pp. 25–48). Boston, MA: Kluwer Academic.

Musqua, D. (1997). *Personal communication,* June 2, University of Saskatchewan, SK.

Nadasdy, P. (1999). The politics of TEK: Power and the "integration" of knowledge. *Arctic Anthropology, 36*(1–2), 1–18.

Nadeau, R., & Désautels, J. (1984). *Epistemology and the teaching of science.* Ottawa, ON: Science Council of Canada.

NASA (National Aeronautics and Space Administration). (2005). *Ancient observations, timeless knowledge.* Author. Retrieved September 14, 2010, from http://sunearthday.nasa.gov/2005/index.htm.

National Indigenous Literacy Association. (2007). *Four directions teaching project.* Retrieved July 20, 2010, from http://www.fourdirectionsteachings.com.

Needham, J. (1956). *Science and civilisation in China* (Vol. 2: History of scientific thought). Cambridge, UK: Cambridge University Press.

Niezen, R. (2003). *The origins of indigenism: Human rights and the politics of identity.* Los Angeles, CA: University of California Press.

Nunavut Bilingual Education Society. (2004). *Taiksumania: Inuit myths and legends.* Iqaluit, NU: Nunavut Bilingual Education Society Publication.

Nunavut Bilingual Education Society. (2006). *Taiksumania: Inuit myths and legends* (vol. 2).

NWT Protected Areas Strategy. (2009). *Protecting natural and cultural areas in the Northwest Territories.* Retrieved July 21, 2010, from http://www.nwtpas.ca/education-trm.asp.

OECD (Organisation for Economic Co-operation and Development). (2006a). *Assessing scientific, reading and mathematical literacy.* Paris, France: OECD.

OECD (Organisation for Economic Co-operation and Development). (2006b). *Evolution of student interest in science and technology studies: Policy report.* Paris, France: OECD.

Office of the Treaty Commissioner. (2009). *Making the connection: Cree First Nations kēhtē-ayak thoughts on education.* Saskatoon, SK: Author.

Ogawa, M. (1995). Science education in a multi-science perspective. *Science Education, 79,* 583–593.

Ogunniyi, M.B. (2007). Teachers' stances and practical arguments regarding a science-Indigenous knowledge curriculum. *International Journal of Science Education, 29,* 963–986.

Orange, A.D. (1981). The beginnings of the British Association, 1831–1851. In R. MacLeod & P. Collins (Eds.), *The parliament of science* (pp. 43–64). Northwood, UK: Science Reviews.

Patchen, R., & Cox-Petersen, A. (2008). Constructing cultural relevance in science: A case study of two elementary teachers. *Science Education, 92*, 994–1014.

Pearson Education Canada. (2010). *Pearson Saskatchewan science 6–8: Program overview.* Toronto, ON: Author.

Peat, F.D. (1994). *Lighting the seventh fire.* New York, NY: Birch Lane Press.

Pedretti, E., & Little, C. (2008). *From engagement to empowerment.* Toronto, ON: Pearson Canada Inc.

Perso, T. (2003). *Improving Aboriginal numeracy.* Perth, Australia: MASTEC, Edith Cowan University.

Pickering, A. (Ed.). (1992). *Science as practice and culture.* Chicago, IL: University of Chicago Press.

Pierotti, R., & Wildcat, D.R. (1997). The science of ecology and Native American tradition. *Winds of Change, 12* (Autumn), 94–97.

Polanyi, J. (2009). The power of reason. Toronto, ON: *Globe and Mail*, May 26, p. A15.

RCAP (Royal Commission on Aboriginal Peoples). (1996). *Gathering strength: Report of the Royal Commission on Aboriginal Peoples.* Ottawa, ON: Canada Communication Group.

Read, T. (2002). *The Kormilda science project: An indigenous perspective on the earth sciences.* Mount Eliza, Australia: Author.

Reiss, M.J. (2008). Should science educators deal with the science/religion issue? *Studies in Science Education, 44*, 157–186.

Reyhner, J. (2006). Dropout nation. *Indian Education Today* (June), 28–30. Retrieved April 27, 2010, from http://jan.ucc.nau.edu/~jar/AIE/IETdropout.html.

Richards, J., Hove, J., & Afolabi, K. (2008). *Understanding the Aboriginal/Non-Aboriginal gap in student performance: Lessons from British Columbia* (Commentary No. 276). Toronto, ON: C.D. Howe Institute.

Richards, J., & Scott, M. (2009). *Aboriginal education: Strengthening the foundations.* Ottawa, ON: Canadian Policy Research Networks.

Riggs, E.M. (2005). Field-based education and indigenous knowledge: Essential components of geoscience education for Native American communities. *Science Education, 89*, 296–313.

Rowe, M.B. (1995, May). Teach your child to wonder. *Reader's Digest*, 177–184.

Rowland, P., & Adkins, C. (2003). Native American science education and its implications for multi-cultural science education. In S.M. Hines (Ed.), *Multicultural science education: Theory, practice, and promise* (pp. 103–120). New York, NY: Peter Lang.

Rudolph, J.L. (2005). Epistemology of the masses: The origins of "the scientific method" in American schools. *History of Education Quarterly, 45*, 341–377.

Ryan, A.G., & Aikenhead, G.S. (1992). Students' preconceptions about the epistemology of science. *Science Education, 76*, 559–580.

Sadar, Z. (1997). Islamic science: The contemporary debate. In H. Selin (Ed.), *Encyclopaedia of the history of science, technology, and medicine in non-western cultures.* Boston, MA: Kluwer Academic.

Saskatchewan Indian Cultural Centre. (2009). *Cultural teachings: First Nations protocols and methodologies.* Saskatoon, SK: Author.

Saskatoon StarPhoenix (2008, January 8). Editorial: Poor literacy of Aboriginals hurting Saskatchewan. Saskatoon, SK.

Schaefer, C. (2006). *Grandmothers counsel the world: Women Elders offer their vision for our planet.* Boston, MA: Trumpeter Books.

Science Council of Canada (SCC). (1984). *Science for every student: Educating Canadians for tomorrow's world* (Report No. 36). Ottawa, ON: SCC.

Science Council of Canada (SCC). (1991). *Northern science for northern society: Building economic self-reliance.* Ottawa, ON: Supply & Services, Government of Canada.

Selin, H. (1992). *Science across cultures: A bibliography on non-Western science and medicine.* New York, NY: Garland.

Semali, L.M., & Kincheloe, J.L. (1999). Introduction: What is indigenous knowledge and why should we study it? In L.M. Semali & J.L. Kincheloe (Eds.), *What is indigenous knowledge? Voices from the academy* (pp. 1–57). New York, NY: Falmer Press.

Settee, P. (2000). The issue of biodiversity, intellectual property rights, and Indigenous rights. In R.G. Laliberte et al. (Eds.), *Expressions in Canadian Native studies* (pp. 459–488). Saskatoon, SK: University of Saskatchewan Extension Press.

Seven Generations Education Institute. (2008). *Bridging the cultural-language gap between renewable energy technology and First Nations communities.* Retrieved September 3, 2010, from http://www.7generations.org/Post%20Secondary%20Pages/Renewable%20Energy%202008.pdf.

Shapin, S. (2010). *Never pure: historical studies of science as if it was produced by people with bodies, situated in time, space, culture, and society, and struggling for credibility and authority.* Baltimore, MD: Johns Hopkins University Press.

Sharpe, A., & Arsenault, J-F. (2009). *Investing in Aboriginal education in Canada: An economic perspective* (CPRN Research Report). Ottawa, ON: Canadian Policy Research Networks.

Sharwood, J., & Khun, M. (2005). *Science Edge 3.* Southbank, Victoria, Australia: Thomson Nelson.

Shope, R. (1998). *We once hunted for buffalo, we now hunt for knowledge: The instructional leadership of Chief Joseph Chasing Horse.* Retrieved September 14, 2010, from http://sunearthday.nasa.gov/2005/na/lakota.htm.

Sillitoe, P. (Ed.) (2007). *Local science vs. global science: Approaches to Indigenous knowledge in international development.* New York, NY: Berghan Books.

Simonelli, R. (1994). Sustainable science: A look at science through historic eyes and through the eyes of indigenous peoples. *Bulletin of Science, Technology & Society, 14,* 1–12.

Simpson, L. (2004). Listening to our ancestors: Rebuilding Indigenous nations in the face of environmental destruction. In J.A. Wainwright (Ed.), *Every grain of sand: Canadian perspectives on ecology and environment* (pp. 121–35). Waterloo, ON: Wilfred Laurier Press.

Snively, G. (1990). Traditional Native Indian beliefs, cultural values, and science instruction. *Canadian Journal of Native Education, 17,* 44–59.

Snively, G., & Corsiglia, J. (2001). Discovering indigenous science: Implications for science education. *Science Education, 85,* 6–34.

Snively, G.J., & Williams, L.B. (2008). "Coming to know": Weaving aboriginal and Western science knowledge, language, and literacy into the science classroom. *L1 – Educational Studies in Language and Literature, 8*(1), 109–133.

Snow, R. (1987). Core concepts for science and technology literacy. *Bulletin of Science Technology Society, 7,* 720–729.

Stanford, P.K. (2006). Instrumentalism. In S. Sarkar & J. Pfeifer (Eds.), *The philosophy of science: An encyclopedia* (Vol. 1, pp. 400–405). New York, NY: Routledge.

Stanley, W.B., & Brickhouse, N.W. (1994). Multiculturalism, universalism, and science education. *Science Education, 78*, 387–398.

Stanley, W.B., & Brickhouse, N.W. (2001). Teaching sciences: The multicultural question revisited. *Science Education, 85*, 35–49.

Statistics Canada (2008). *Aboriginal peoples in Canada in 2006: Inuit, Métis and First Nations census* (Release no. 5: January 15). Retrieved September 14, 2010, from http://www12.statcan.ca/english/census06/release/aboriginal.cfm.

Stewart, G. (2005). Mäori in the science curriculum: Developments and possibilities. *Educational Philosophy and Theory, 37*, 851–870.

Stonechild, B. (2007). *The new buffalo: The struggle for Aboriginal post-secondary education in Canada*. Winnipeg, MB: University of Manitoba Press.

Sutherland, D.L. (2005). Resiliency and collateral learning in science in some students of Cree ancestry. *Science Education, 89*, 595–613.

Sutherland, D., & Dennick, R. (2002). Exploring culture, language and the perception of the nature of science. *International Journal of Science Education, 24*, 1–25.

Sutherland, D., & Henning, D. (2009). *Ininiwi-kisk n tamowin*: A framework for long-term science education. *Canadian Journal of Science, Mathematics and Technology Education, 9*, 173–190.

Sutherland, D., & Tays, N. (2004, April). *Incorporating indigenous culture into school science*. Paper presented at the annual meeting of the National Association for Research in Science Teaching, Vancouver, BC. Available from d.sutherland@uwinnipeg.ca.

Suzuki, D. (2006). *David Suzuki: The autobiography*. Vancouver, BC: Greystone Books.

Taconis, R., & Kessels, U. (2009). How choosing science depends on students' individual fit to 'science culture.' *International Journal of Science Education, 31*, 1115–1132.

Tang, Z., Coffey, J.E., Elby, A, & Levin, D.M. (2010). The scientific method and scientific inquiry: Tensions in teaching and learning. *Science Education, 94*, 29–47.

Taylor, A., Friedel, T.L., & Edge, L. (2009). *Pathways for First Nation and Métis youth in the oil sands* (CPRN Research Report). Ottawa, ON: Canadian Policy Research Networks.

Turner, N.J., Ignace, M.B., & Ignace, R. (2000). Traditional ecological knowledge and wisdom of Aboriginal peoples in British Columbia. *Ecological Applications, 10*, 1275–1287.

van Eijck, M., & Roth, W-M. (2007). Keeping the local local: Recalibrating the status of science and Traditional Ecological Knowledge (TEK) in education. *Science Education, 91*, 926–947.

Viergever, M. (1999). Indigenous knowledge: An interpretation of views from indigenous peoples. In L.M. Semali & J.L. Kincheloe (Eds.), *What is indigenous knowledge? Voices from the academy* (pp. 333–359). New York, NY: Falmer Press.

Wagamese, R. (2008). *One Native life*. Toronto, ON: Douglas & McIntyre.

Watson, H., & Chambers, D.W. (1989). *Singing the land, signing the land*. Geelong, Victoria, Australia: Deakin University Press.

Windschitl, M., Thompson, J., & Braaten, M. (2008). Beyond the scientific method: Model-based inquiry as a new paradigm of preference for school science investigations. *Science Education, 92*, 941–967.

Wood, A., & Lewthwaite, B. (2008). Mäori science education in Aotearoa New Zealand: He pütea whakarawe: Aspirations and realities. *Cultural Studies of Science Education, 3*, 625–654.

Wong, S.L., & Hodson. D. (2009). From the horse's mouth: What scientists say about scientific investigation and scientific knowledge. *Science Education, 93*, 109–130.

Wong, S.L., & Hodson. D. (2010). More from the Horse's Mouth: What Scientists Say about Science as a Social Practice. *International Journal in Science Education, 32*, 1431–1463.

Worster, D. (1994). *Nature's economy: A history of ecological ideas.* Cambridge, UK: Cambridge University Press.

Yazzie, R. (1996, July). *Law as a form of cultural restoration and healing.* A paper presented to the International SSHRCC Summer Institute on Cultural Restoration of Oppressed Indigenous Peoples, University of Saskatchewan, Saskatoon, SK.

Yore, L.D., Hand, B.M., & Florence, M.K. (2004). Scientists' views of science, models of writing, and science writing practices. *Journal of Research in Science Teaching, 41*, 338–369.

Ziman, J. (1984). *An introduction to science studies: The philosophical and social aspects of science and technology.* Cambridge, UK: Cambridge University Press.

Ziman, J. (2000). *Real science: What it is and what it means.* Cambridge, UK: Cambridge University Press.

Zukav, G. (1979). *The dancing Wu Li masters: An overview of the new physics.* New York, NY: Bantam Books.

Zwick, T.T., & Miller, K.W. (1996). A comparison of integrated outdoor education activities and traditional science learning with American Indian students. *Journal of American Indian Education, 35*(2), 1–9.

INDEX

Alaska Native Knowledge Network (ANKN), 15, 170
Algonkin (Algonkian Nation), ix
alienation, *see* marginalization
American Association for the Advancement of Science (AAAS), 9, 22, 24
anthropocentrism, 52, 61, 111, 115, 163
Aotearoa New Zealand, xi, 2–3, 16, 71, 154
argumentation, *see* scientific arguments-persuasion
assessment, 39, 83, 95, 125, 153
    see also students
assimilation, 8, 14, 35, 64, 142
Australia, xii, 2, 3, 16–17, 20 31, 80–81, 171–172, 174
Australian Academy of Science, 154, 171
axiology, 113

balance (harmony), 7, 9, 69, 74, 77–79, 80–82, 86–87, 90, 101–102, 104, 109, 111, 117, 120, 137–139, 143–144, 146, 148, 165
Barnhardt, Ray, 15, 27, 70, 79, 83, 84, 86, 88, 89, 122, 134, 144, 165, 170
Battiste, Marie, 6, 8, 12, 47, 64, 75, 77, 83, 89, 90, 94, 97, 104, 142
Bear, Judy, 27, 154
Belczewski, Andrea, 18, 26, 92, 97, 155
believing *versus* understanding, 94, 101, 135, 164
biodiversity, 12
biopiracy, 11, 13, 25
Blackfoot, 67, 75, 87, 93, 175
Blackwater, Elder Andy, 51, 93
British Association for the Advancement of Science (BAAS), 21–22, 31
building cultural bridges, 5, 6, 14, 18, 30, 43, 54, 62, 63, 68, 71, 73, 91, 97, 121–155

Cajete, Greg, 3, 9, 12, 18, 27, 31, 52, 65, 67, 68–69, 73, 74, 75, 76, 78, 79, 81, 85, 86, 88, 89, 90, 93, 95, 105, 107, 131, 142, 155, 160, 162, 164, 165, 173

Canadian Council on Learning, 7, 8, 27, 98, 132, 165, 169
careers (occupations), science-based, x, 7–8, 9, 11–13, 18, 19, 23, 24–25, 38, 40, 44–45, 48, 61, 128, 134, 149, 159, 167
Cartesian dualism, *see* dualism
ceremony, 69, 71, 73, 77, 79–80, 85, 87–88, 92, 95, 101, 108, 125–126, 137, 141, 173, 175
Chasing Horse, Chief Joseph, 10
Chickasaw, 92, 94
classroom environment, 6, 102, 133–135, 148, 154
coexistence, 5, 18, 61, 73, 82, 89, 91, 97, 99, 102, 110, 111, 113, 114, 118, 141, 146, 147, 158
collateral learning, *see* learning
colonization, 2, 9, 13, 20, 25, 54, 64–67, 81, 97, 103, 111–112, 120, 122, 148, 162, 173
coming to know, *see* learning
communal, 38, 43, 60, 106–107, 110, 118, 131
communities of practice (scientific), 24, 28, 42, 60
community-based, 68, 84, 109, 123, 131, 144, 154
    instruction, *see* instructional approaches
    resources, x, 123–129, 133–135
comparisons of Indigenous ways of living in nature and Eurocentric sciences, 6, 8, 43–44, 48, 49, 50, 51, 52, 65, 66, 68–69, 70, 71, 73, 75, 76, 77, 78, 79, 81–82, 83, 84, 85–86, 87, 88, 89–90, 91, 92, 95, 97, 99–120, 132, 148, 163
complementary, 5, 18, 35, 61, 89, 91, 99, 102, 102, 110, 132, 140
concepts of evidence, 40–41
conflict, 96, 108
    see also culture (clash)
consensus (consensus making), 23, 34–35, 39, 41, 55, 57, 61, 90, 96, 106, 118, 134
constructivism, *see* social constructivism
Council of Ministers of Education Canada, 2, 9

Cree, x, 12, 17, 27, 31, 67, 69–70, 73–74, 76, 78–79, 81–82, 88, 93, 95, 103–105, 123, 155, 162, 174
critical realism, *see* realism
cross-cultural, 30, 68, 130
    see also school science
cultural differences:
    students and school science, 1, 7–9, 18, 46, 49, 52, 79, 100–103, 141–142
    Indigenous and scientific ways of knowing nature, 108–112
cultural-historical activity theory, 57, 61
cultural responsiveness, 71, 104, 128–129
    see also school science
cultural similarities:
    students and school science, 1, 9, 12–13, 48–49, 141, 142, 163
    Indigenous and scientific ways of knowing nature, 106–108
culture:
    clash, 8, 91, 101–103, 122, 125, 141–142, 150–153
    culture of Eurocentric science, 7, 19, 24, 28–29, 33, 43–59, 102, 105, 141, 156–157
    defined, 2
culture-free, 57, 59, 108, 161
culture-laden, 59, 109
curiosity, 24, 38, 44–45, 49, 82, 119–120
curriculum, *see* school science (conventional, and enhanced curriculum)

Dakota, 88
decolonization, 97
Dëne, x, 67, 87
dichotomy, 5, 28–29, 52, 75–76, 110, 112
Diné (Navajo), 67, 74, 78, 100–103, 141
disciplinary matrix, 34
dualism, Cartesian, 50–51, 61,75, 80, 92, 108, 110, 115, 117
Duran, Phillip, 12, 51
Dyck, Senator Lillian, 12, 44, 51, 58, 85, 107
dynamic knowledge, *see* knowledge

ecology, 6, 11–12, 14, 24, 34, 51–52, 65, 73–75, 77, 84, 90, 127, 142, 153, 168, 173
economy, 8, 10, 13–14, 19, 25, 39, 45, 59, 64, 108, 112, 118, 142, 159–160, 172
Einstein, Albert, 12, 52–53, 55–56, 58
Elders, 3, 5, 6, 11, 16–17, 25, 65, 69–70, 73, 75–76, 78–80, 83–86, 88, 90–91, 93, 95, 104, 106–109, 113, 118, 123–126, 129–130, 133, 135, 141–142, 144, 146, 153–154, 168, 175
empirical (empiricism), 5, 20, 23, 31, 35, 38–42, 46, 48, 56–57, 61, 69, 82–86, 91, 95, 107–108, 116, 162
enculturation, 29
engineers, 7, 12–13, 21, 49, 55–56, 83–84, 128, 133–134, 142, 161, 169
environment, 9, 11–12, 14–15, 24, 34, 36, 39, 66, 69, 74, 76–77, 79, 88–89, 96–98, 109–110, 114, 117,118, 139–140, 154, 167, 172, 175
epistemology, 17, 42, 70, 77–78, 113, 162
equity, 30–31, 104, 143, 145
    equity and social justice, 7–9
Ermine, Willy, 69, 74, 76–77, 79, 81, 83, 87, 92
ethics, 8, 12, 39–41, 76, 95, 108, 111, 119, 125, 138, 140, 163–164
Euclidean geometry, 53–54, 56, 86, 90
Euro-American, 5, 22, 28–29, 31, 44, 53–54, 59, 63–65, 68, 76, 161
Eurocentric science, *see* science
experiments, 5, 28, 33, 36–38, 46, 48, 51–52, 57–58, 82–84, 107–108, 117, 146, 151, 153, 157
    experimental strategies, 38–39

field studies, 21–22, 38, 116
First Nations, defined, 2
flux, 81, 83, 87, 94, 110, 115, 119, 146
Forests and Oceans for the Future project, 17, 170
formative assessment, 144
Four Arrows, 80, 92
Four Directions Teaching project, 72
Fulbright, Senator J. William, 69

generalizability, 21, 30, 34, 44, 48–49, 61, 72–75, 86, 103, 111, 116, 156
global contexts, 14, 25, 41, 172
global science, 30
globalization, 11, 13, 23, 25, 30, 111–112, 114

harmony, *see* balance
hierarchy, 52, 78
Hmong, 149–153
holism, 12, 27, 51, 69, 73, 76–79, 84–85, 87, 102, 108–109, 111, 116, 125, 131–132, 137, 142, 145, 163–164, 169, 173

humility, 10, 78, 81, 97, 124, 133, 142
hybrid, hybridized (knowledge, learning, and understanding), 14, 144, 154
hypothesis, scientific, 36, 38–39, 44, 46, 59, 116, 146

idealized, 41, 43, 53, 55, 58, 60, 69, 105, 111, 133
    see also knowledge
identity, ix, 47, 64, 74, 89, 120, 141
    see also students (self-identity)
ideology, 12, 22, 42–43, 45, 52, 57–58, 101, 111, 165
    defined, 43
incommensurate, 34–35, 96
Indigenous:
    beliefs, 14, 27, 48, 85, 93, 101–102, 115, 125, 128, 142, 150, 152
    community, xii, 3–4, 6, 8, 13, 16, 52, 66–69, 71–72, 83, 124, 126–130, 138–139, 144, 146, 148, 153–154
    defined, 2, 63–65, 67
    knowledge fairs, 127, 140–141, 144
    languages, see language
Indigenous Education Institute, 16
Indigenous ways of living in nature (IWLN), defined, 65
inner space, 76–77, 79, 83, 87, 92
instructional approaches (teaching), 136–141
    community-based, 123–129, 133–135
    learner-centred, 140
    culturally responsive, see school science
instrumentalism, 54, 56–57, 61, 161
intellectual tradition, 109, 110, 116
intelligible essences, 47, 53, 55, 89, 111, 118
International Council for Science, 9, 10
intuition, 1, 5, 12, 41, 47, 50, 58, 60, 76, 85, 91–92, 97, 107, 110, 116, 119, 124, 131, 143
Inuit, x, xii, 17, 25, 27, 63, 66–67, 96, 130, 132, 154, 168–169
    defined, 2
Inuit Qaujimajatuqangit, 17, 96, 169
Inuit Subject Advisory Committee, 17
Inuuqatigiit, 17
invisible colleges, 23, 33, 38, 118

Kawagley, Oscar, 3, 6, 15, 31, 70–71, 75–76, 79, 83–84, 86, 88–89, 127, 144

knowledge:
    dynamic (tentative), 47–48, 57, 61, 65, 77, 82–83, 107, 116
    Eurocentric science perspective, 43–62, 106–120
    idealized, 36, 48–49, 55, 61, 151
    Indigenous perspective, 65–69, 94–98, 106–120
Knowledge Keeper, 27, 65, 68, 72, 80, 85, 91, 123–126, 153–154, 156, 170
Kuhn, Thomas, 33–35, 37–38, 58, 61

Lakota, 10, 67, 78, 89, 91, 92
language:
    English, 22, 63, 65, 68, 103–105
    French, 63, 103–105
    Indigenous, x, 3, 6, 16, 18, 65–68, 71, 75, 84, 101–102, 103–105, 107, 118, 126–127, 129–130, 135, 145–148, 153, 155, 162, 163, 168, 175
    scientific, 8, 19, 28–29, 44, 47, 53, 55, 76, 103, 107, 118, 136
law, scientific, 21, 23, 44, 46–48, 54–57, 59, 109
learning:
    classroom environment, see classroom environment
    collateral learning, 114
    coming to know, 69–71, 94, 99, 131–132, 138, 144, 153, 162
    experiential learning, xii, 70, 97–98, 131, 138, 142, 153, 155
    hybridized learning, 14, 114, 154
    Indigenous student learning, 131–132
    lifelong learning, 27, 89, 132, 153–154, 156
    recurrent learning strengths, 131–132, 137, 138–139, 165
    resistance to learning, 8–9, 17, 46, 49, 52, 64, 79, 80, 120, 143, 147
    two-way learning, 14, 142
learning styles, see learning (recurrent learning strengths)
Lickers, Henry, 11, 98
lifelong learning, see learning
literacy:
    ecological (environmental), 15, 89, 153, 165
    nature, 10, 155, 165
    oral, 82, 165
    scientific, xi, 9, 10, 18, 59, 165
Little Bear, Elder Leroy, 75, 78, 87, 146

Loo, Seng Piew, 31, 51, 75
lost in translation, 97, 104, 163

MacIvor, Madeleine, 8, 13–14, 75, 76, 133, 137, 164
Māori, 2–3, 16, 31, 71, 154
marginalization, ix, 5, 8–9, 18, 64, 101, 112, 143
Marshall, Elder Albert, 113, 142
Maryboy, Nancy, 67, 69, 74, 130
mathematics, 28, 38, 49, 80–81, 86, 90–91, 107, 137, 156–157, 169
   see also quantification
McKinley, Elizabeth, 3, 9, 13, 16, 30, 31, 64, 145, 161, 162, 166
media, 5, 10, 26, 29, 32, 36, 38, 55, 59–60, 139
Medicine Wheel, 107
metaphorical, 44, 54, 55, 61, 76–77, 81, 89–90, 93, 107, 108, 116, 162, 173
metaphysics, 28, 50, 59, 75, 76–77, 92, 95, 115
Métis, x, xi, xii, 13, 24, 25, 27, 67, 74, 75, 132, 133, 135, 140, 144, 156–157
   defined, 2
Mi'kmaq (Mi'kmaw Nation), 6, 67, 94, 142
Ministries of Education (Departments of Education), xi, 3, 4, 5, 7, 8, 9, 15, 158
models:
   cognitive, 24, 34, 38, 40, 44, 46, 54–57, 59, 71, 89, 107, 117, 137, 138
   heuristic, 27, 37, 44, 56–57, 71, 132, 169
   learning, 130, 142
   role models, 22, 128–129, 133, 156, 169
Mohawk, 11, 67
monism (monist), 51, 75–77, 84–85, 92, 108, 110, 115, 116, 117, 164
Mother Earth, 66–67, 73–74, 76, 78, 82, 98, 107, 109, 110–111, 113, 115, 117, 119, 125, 154, 163, 175
multicultural classrooms, settings, xii, 15, 31, 32, 67, 128, 133, 134, 145, 153, 155
Musqua, Elder Danny, 85, 87, 139
mystery, 44, 60, 81–82, 97, 108, 110–111, 115, 118, 136, 171
myths (stories):
   Indigenous, 71, 73, 74, 81–82, 87, 90, 93, 96, 173, 101, 102, 107, 117, 130, 135, 153, 175
   scientific, 26, 36–37, 41–42, 54, 57, 59, 174

Nakawē (Saulteaux), 85, 87, 139
natural philosophy (natural philosophers), 20–22, 23, 37, 44, 48, 52, 76, 108, 111, 118, 160
nature:
   Eurocentric science perspective on, 44–46, 50–54, 66, 68, 75
   Indigenous perspective on, 65–67, 68, 75
Nature of science (NOS), 13, 41, 105
Navajo, *see* Diné
neo-colonial (neo-colonialism), 112, 122
neo-indigenous, 31–32, 65
New Zealand, *see* Aotearoa New Zealand
Nisga'a, 106, 122
non-verbal communication, 148
Nunavut Bilingual Education Society, 96

objectify, 53
objectivity, 41–43, 52–53, 55, 57–58, 59, 70, 79, 81, 95, 111, 115, 116, 119, 120, 161
Ogawa, Masakata, 30, 31, 32, 160
Ojibwa, 67, 80, 81, 92, 97, 147, 176
Omushkego Oral History project, 72
Oneida, 67
ontology, 113
outer space, 76, 81, 83, 92

paradigm, 33–36, 37, 38, 42, 43–44, 46, 48, 50–51, 53, 55–56, 57–58, 59, 61, 77, 107, 109, 116, 118, 164
Peru, 2
place-based, 16, 70, 72–75, 86, 99, 111, 114, 116, 118, 123, 130, 146, 155, 173
pluralism, 31–32, 43, 74, 83, 89, 90, 99, 105, 155, 160, 164
Polanyi, John, 23, 47, 55, 61
positivism, 57–59, 61, 90, 105, 141, 161–162
postcolonial, 97, 123
power and dominion (control) over nature, 20, 30, 38, 42, 52, 58, 78, 83, 84, 97, 108, 111, 119, 142
professional development, teachers, 4, 15, 17, 159, 169
protocol, 68, 70, 72, 79, 80, 82, 93, 95, 101, 108, 123–126, 129, 144, 156
pseudoscience versus science, 37–38
Pueblo, 12, 65, 73, 95, 171

quality-control monitoring procedures, 38, 39–40, 46, 117

quantification, 44, 50, 52–54, 61, 81, 117
Quechua, 2, 174

racism, 8, 103, 112, 134
R&D (research and development), 12, 22, 23–24, 25, 26, 39, 45, 160
rational, 5, 20, 27, 30, 38, 42, 57, 58, 61, 82, 90–91, 92, 93, 106, 107, 109, 116, 150, 160, 164
realism (critical realism), 54–57, 58, 59, 61, 81, 90, 161
recurrent learning strengths, *see* learning
reductionism, 51, 61, 77, 108, 111, 116
reification (reify), 54, 55
rekindling traditions, x, 17, 94, 128, 133, 156–158, 170
relational, 52, 66, 68–69, 70, 73–74, 77, 78–81, 82, 84, 85, 86, 87, 92, 94–95, 96, 98, 99, 102, 109, 110, 111, 115, 116, 117, 118, 119, 123–124, 125, 128–129, 131, 132, 133, 135, 138, 142, 144, 146, 154, 155, 163, 175
relativism, 31
reliability, 30, 39, 40, 57, 82, 106, 157
resiliency, 14, 120, 126, 173
resistance, 80, 120, 147, 173
  see also learning
resource management, 7–8, 11, 12, 13, 25, 38, 68, 83, 99, 110, 122, 128, 139, 142, 163, 170
resources for science teachers, xii, 4, 29, 32, 55, 71, 121, 123–130, 133, 134, 137, 140, 141, 142, 143, 146, 158, 167–172
respect, 3, 5, 9, 14, 15, 32, 68, 74, 79, 80, 92, 93, 94, 96, 102, 106, 109, 111, 119, 123, 125, 128, 129, 133, 134, 135, 140, 142, 148, 151, 152, 158, 165, 170
responsibility in relationships, ix, 66, 69, 78–80, 84, 86, 92, 115, 119, 138, 144, 155
Riemannian geometry, 53, 90
role models, 22, 128–129, 133, 156, 169
Royal Commission on Aboriginal Peoples, 8, 14
rural settings, xii, 64, 66, 70, 71, 128

Sámi, 2
Saskatchewan Indian Cultural Centre, 80, 85, 92, 93, 97, 123, 126
Saulteaux, *see* Nakawē
Schaefer, Carol, 78, 85, 113, 175
school science:
  budget, 139, 140
  conventional, 3, 9, 26, 32, 46, 49, 59–60, 71, 105, 132, 144, 163, 165
  cross-cultural, x, 5–6, 15–18, 94, 100, 112, 114, 123, 129–130, 136, 142, 153–155, 156–158, 165, 170, 173
  culturally responsive, 3–4, 9, 15–17, 31, 121–155, 165, 170
  enhanced curriculum, 4, 7, 9, 10, 12, 13–14, 15–18, 19, 70, 71, 75, 97, 123, 124, 126, 129, 155, 159, 163, 171
  multicultural, xii, 15, 31, 32, 67, 128, 133, 134, 145, 153, 155
science:
  academic science, 23–24, 25, 33, 36, 48, 49, 60, 162
  core science, 42, 61, 162
  corporate science, 11, 22–23, 24–25, 28, 36, 38, 39, 45, 118, 122
  curriculum, *see* school science (conventional, and enhanced curriculum)
  defined, 30–32
  ethnoscience, 28–29
  Eurocentric science defined, 4, 19, 22, 28, 30–32, 37–38, 60
  Eurocentric sciences (ES), defined, 35
  frontier science, 42, 61, 162
  global science, 30
  pluralist science, *see* pluralism
  post-academic science, 23–24, 25, 41, 160
  "pure" science, 23, 25, 44–45
Science Council of Canada, 7, 9
science-based:
  activities, 16, 37, 41, 125, 127, 129, 132, 133, 138, 140, 144, 146, 149, 157
  news items, 5, 10, 26, 29, 32, 36, 38, 55, 59–60, 138, 139
  occupations, *see* careers
science-technology-society environment (STSE), 2, 26, 41, 105
scientific arguments-persuasion, 34, 39, 41, 55, 56–57, 61, 101, 118
scientific beliefs, 19, 27–29, 34–35, 43, 47, 55, 58, 60, 115
scientific literacy, *see* literacy
scientific method, the, 36–37, 42, 57, 59
scientific methods, 37, 39, 67, 59, 95, 108

scientists:
    descriptions of, 10–12, 19, 22–26, 28–29, 33–36, 37, 39, 41, 42–43, 47, 50, 51, 53, 54, 56, 57, 58, 59–60, 61, 70, 75, 83, 85, 86, 91, 106–120
    empirical, 38–39
    social goals, 44–46, 61, 110, 118
    theoretical, 38, 45, 52–53
self-identity, *see* students
Seneca, 11, 98
sensitivity, 71, 95, 103–105, 128, 129, 131
similarities, *see* cultural similarities
Snively, Gloria, 10, 11, 17, 73, 84, 85, 89, 90, 99, 106, 109, 114, 142
snowshoes, 3, 71, 126, 127, 140, 147, 156–158
social constructivism, 57, 61, 161
socio-political context, 111, 112, 121–122
socio-scientific issues (SSI), 41, 105
sovereignty, 2, 13–14
spirituality, 14, 34, 50, 53, 66, 69, 70, 71, 73, 74, 75–77, 78–80, 81, 83, 84–85, 87, 88, 89, 91, 92–94, 95, 97, 101, 107, 108, 109, 110, 111, 113, 115, 116, 117, 122, 126, 131, 135, 142, 144, 164, 165, 175
stereotyping, 4–5, 6, 19, 27, 36, 59–60, 63, 72, 103, 112, 129, 131, 139
stewardship, 14, 42, 96, 119, 126, 163
students:
    achievement, 2, 11, 15–18, 102, 134, 142, 143, 148–149, 154, 169, 170
    assessment, 68, 70, 133, 136, 143–146, 157, 164
    Indigenous, 2, 8–9, 18, 52, 61, 67, 68, 71, 100–103, 104, 112, 127, 128, 129, 131–133, 134, 135, 136, 137–138, 139–140, 142, 148–149, 154, 155, 161, 162, 163
    non-Indigenous, xii, 4, 12, 14, 16, 18, 52, 97, 112, 127, 132, 133, 135, 136, 139, 141, 156, 167, 162, 164, 165
    reaction to school science, *see* learning (resistance to), and students (achievement)
    science-oriented, 1, 9, 12–13, 48, 49, 132, 141
    science-shy, 1, 46, 49, 52
    self-identity (cultural identity), xii, 9, 69, 71, 74, 94, 101–102, 120, 136, 142, 147, 151
subjectivity, 33, 35, 39, 41–43, 50, 57–58, 60, 61, 85, 111, 115, 116, 119, 161, 162

Sundance ceremony, 87
superstition, 91
sustainability, x, 9, 10–12, 14, 23, 42, 51, 73, 76, 79, 88, 114, 120, 126, 155, 165, 170, 172, 175
Sutherland, Dawn, 17, 31, 126, 138, 153
Suzuki, David, 10, 72, 76, 85, 87, 89, 94, 112, 164

Taiwan, xii, 2
teacher candidates, xi, 4
teacher expectations, 101, 132, 139, 141, 143–144, 145, 148–149, 154
teaching, *see* instructional approaches
teaching materials and resources, 4, 15, 17, 29, 32, 55, 106, 112, 123–128, 129–130, 133–135, 137, 139–140, 156–158, 169, 170
technology, 9, 14, 19, 20, 21, 26, 28, 49–50, 68, 71, 73, 80, 83, 95, 89, 96, 97, 102, 106, 107, 108, 117, 125, 126, 129, 140, 157, 171
    engineers, *see* engineers
    defined, 21, 26
    technicians, 7, 11, 12, 13, 21, 117, 134, 142
    technologists, 21, 22
tentative knowledge, *see* knowledge (dynamic)
textbooks, 16, 58, 124, 125, 129, 136, 143–144, 158
theory, scientific, 12–13, 24, 38, 44, 46, 51, 53, 58, 135, 174
theory-laden, 35, 43, 54, 60, 81
Thunderbird, 80, 93
time:
    cyclical, 50, 86–88, 109, 111, 118
    rectilinear, 49–50, 61, 87, 88, 109, 111, 118
tobacco, 80, 123
traditional ecological knowledge and wisdom (TEKW), 11, 24, 84
transformer, 81, 136, 175
trickster, *see* transformer
truth, 13, 20, 21, 41, 47, 55, 61, 89–90, 92, 109, 116
two-eyed seeing, 113, 114, 142
two-way:
    learning, 14
    approaches, 5, 142

under-representation, 7–9, 18, 61, 161
understanding versus believing, *see* believing versus understanding

United States, xii, 2, 3, 15–16, 22, 64, 67, 137, 150, 170–171
universalism, 29–30, 31, 32, 43, 48, 56, 57–58, 60, 99, 101, 141, 161
urban settings, xii, 4, 31, 66–67, 70, 71, 128, 133–134, 148

validity:
    content, 47, 56, 88–89, 111, 118
    cultural, 15, 72, 88, 147
    predictive, 46–47, 56, 61, 89, 111, 118
value-free, 57, 59
value-laden, 39, 42, 43, 162
values (value systems), 8, 12, 19, 28, 37, 39, 41–43, 59, 67, 68, 78, 79, 81, 88, 93, 106–107, 109, 111, 119, 124, 127, 129, 144, 152, 156, 163, 168
    types of values in the scientific enterprise, 34, 39, 41–42, 44, 52, 57, 58, 60, 106–107, 111, 139, 156, 162

Wagamese, Richard, 92, 97, 134, 147, 148, 176
walk in both worlds, 141

Walpole First Nation, 24
Wilson, Lee, 24
wisdom (wisdom-in-action), 68, 69–70, 76, 79, 84, 94, 97, 106, 107, 109, 110, 117, 118, 123, 124, 125, 131, 137, 140, 142, 154, 155, 163, 174, 175
    teachers' inner wisdom, 121
wisdom tradition, 109, 110, 116
worldview:
    defined, 1, 26–27
    Euclidean, 53, 86
    Eurocentric (Euro-American), 28, 59, 87, 160–161
    Indigenous (defined, 73), 2, 27, 63, 66, 67, 68, 71, 75, 78, 81, 88, 95, 96, 97–98, 99, 100, 102, 104, 108, 123, 126, 128, 132, 135, 141, 142, 145, 146, 163, 173–176
    scientific (defined, 43), 1, 18, 27–28, 43, 48, 58, 59

yin and yang, 99, 147
Yupiaq, 6, 15, 31, 71, 75, 79, 84

Ziman, John, 23, 24, 25, 26, 38, 39, 41, 45, 51, 55, 56, 57, 59, 61, 161, 162